# THE HUMAN NETWORK

'Compelling... Professor Jackson manages not only to present a lot of complex research engagingly but to show how the key concepts of network theory relate to a wide range of contemporary issues, from financial contagions to the spread of fake news.'

*New York Times*

'This thought-provoking book explains how and why human networks matter so much, and why they can be a source of strength and fragility. A must-read.'

Daron Acemoglu, bestselling author of *Why Nations Fail*

'Spanning a rich array of topics, including the spread of epidemics and financial crises, political polarization, and economic inequality, *The Human Network* is a highly readable yet deeply informed survey of social life viewed through the lens of networks.'

Duncan Watts, author of *Six Degrees: The Science of a Connected Age*

'Very engaging and worthwhile.'

Diane Coyle, Professor of Public Policy, University of Cambridge

## ABOUT THE AUTHOR

Matthew O. Jackson is the William D. Eberle Professor of Economics at Stanford University (where he received his PhD in 1988), an external faculty member of the Santa Fe Institute, and a fellow of the Canadian Institute for Advanced Research. He is a member of the National Academy of Sciences, a fellow of the American Academy of Sciences, a fellow of the Econometric Society, a Game Theory Society Fellow, and an Economic Theory Fellow. He has received the Social Choice and Welfare Prize, the John von Neumann Award from Rajk Laszlo College, the Berkeley Electronic Press Arrow Prize for Senior Economists, and a Guggenheim Fellowship. He teaches an online course on social and economic networks, and co-teaches (with Kevin Leyton-Brown and Yoav Shoham) two online game-theory courses, which together have reached more than a million students. He is the author of *Social and Economic Networks* and *Handbook of Social Economics*.

# THE
# HUMAN
# NETWORK

## How We're Connected and Why It Matters

# MATTHEW O. JACKSON

Atlantic Books
London

First published in the United States in 2019 by Pantheon Books, a division of
Penguin Random House LLC, New York.

First published in trade paperback in Great Britain in 2019 by Atlantic Books,
an imprint of Atlantic Books Ltd.

This edition published in 2020.

Grateful acknowledgment is made for permission to reprint images on the
following pages: Page 62: Barbulat/Shutterstock.com; Page 83: Reprinted from
*Physica A: Statistical Mechanics and Its Applications*, volume 379, "The Topology
of Interbank Payment Flows," by Kimmo Soramäki, Morten L. Bech, Jeffrey
Arnold, Robert J. Glass, and Walter E. Beyeler, pages 317–33. Copyright © 2007
by Elsevier B.V. Reprinted by permission of Elsevier B.V.

All other images created by the author.

Every effort has been made to trace or contact all copyright holders. The
publishers will be pleased to make good any omissions or rectify any mistakes
brought to their attention at the earliest opportunity.

10 9 8 7 6 5 4 3 2

A CIP catalogue record for this book is available from the British Library.

Paperback ISBN: 978 1 78649 022 3
E-book ISBN: 978 1 78649 021 6

Printed in Great Britain

Atlantic Books
An imprint of Atlantic Books Ltd
Ormond House
26–27 Boswell Street
London
WC1N 3JZ

www.atlantic-books.co.uk

*For Sally and Hal*

# CONTENTS

# CONTENTS

# THE HUMAN NETWORK

# 1 · INTRODUCTION:
# NETWORKS AND HUMAN BEHAVIOR

## The More Things Change

*"In Globalization 1.0, which began around 1492, the world went from size large to size medium. In Globalization 2.0, the era that introduced us to multinational companies, it went from size medium to size small. And then around 2000 came Globalization 3.0, in which the world went from being small to tiny."*

—THOMAS FRIEDMAN, INTERVIEW IN *WIRED*
(AUTHOR OF *THE WORLD IS FLAT*)

On December 17, 2010, Mohamed Bouazizi, a twenty-six-year-old street vendor in the dusty small city of Sidi Bouzid in central Tunisia, lit himself on fire. He did so as a desperate statement of outrage at the tyrannical government that had ruled Tunisia for more than two decades and repeatedly crushed any opposition. His family had long been outspoken against the government and he found himself regularly harassed by the local police. That morning, the police publicly humiliated him and confiscated his day's produce. Mohamed had borrowed the money to buy his produce, and its loss was the last of many straws. Mohamed drenched himself in gasoline and burned himself alive in protest.

Decades ago, the several-thousand-person protest that quickly followed would have been the end of the story. Few outside of Sidi Bouzid would have even been aware that anything happened. However, videos of the aftermath of Mohamed Bouazizi's self-immolation were impossible to contain and were quickly shared via social media and reported widely. News of the Tunisian and other governments' oppression had already been spreading after confidential documents appeared weeks earlier on WikiLeaks. The Arab Spring that would follow was enabled by and coordinated via social media such as Facebook and Twitter as well as cell phones.[1]

Although the methods of communication were modern, ultimately it was a network of humans spreading news and outrage. What was new was how widely and quickly news could spread, and how people were able to coordinate their responses. But understanding what happened still boils down to understanding how news spreads between people and how their behaviors influence each other.

The size and ferocity of the resulting Tunisian protests toppled the government by mid-January. The insurgency had also spread to neighboring Algeria, and over the next two months erupted in Oman, Egypt, Yemen, Bahrain, Kuwait, Libya, Morocco, and Syria, and even Saudi Arabia, Qatar, and the United Arab Emirates. The successes and failures of the Arab Spring are open to debate. But the swift proliferation of protests throughout that part of the world was not only unprecedented but highlighted the importance of human networks in our lives.

As dramatic as recent changes in human communication have been, as Thomas Friedman's quote above indicates, the world has shrunk many times before—in the wake of: the printing press, the posting of letters, overseas travel, trains, the telegraph, the telephone, the radio, airplanes, television, and the fax machine. Internet technology and social media are only the latest chapter in the long history of changes in how people interact, at what distance, how quickly, and with whom.

Yet even as networks of interactions between humans change, much about them is enduring and predictable. Understanding human networks, as well as how they are changing, can help us to answer many questions about our world, such as: How does a person's position in a network determine their influence and power? What systematic errors do we make when forming opinions based on what we learn from our friends? How do financial contagions work and why are they different from the spread of a flu? How do splits in our social networks feed inequality, immobility, and polarization? How is globalization changing international conflict and wars?

Despite their prominent role in the answers to these questions, human networks are often overlooked when people analyze important political and economic behaviors and trends. This is not to say that we have not been studying networks, but instead that there is a chasm between our scientific knowledge of networks as drivers of human behavior and what the general public and policymakers know. This book is meant to help close that gap.

Each chapter shows how accounting for networks of human relationships changes our thinking about an issue. Thus, the theme of this book is how networks enhance our understanding of many of our social and economic behaviors.

There are a few key patterns of networks that matter, and so the story here involves more than just one idea hammered home. By the end of this book, you should be more keenly aware of the importance of several aspects of the networks in which you live. Our discussion will also involve two different perspectives: one is how networks form and why they exhibit certain key patterns, and the other is how those patterns determine our power, opinions, opportunities, behaviors, and accomplishments.

## Billions Upon Billions of Networks

*"Life is really simple, but we insist on making it complicated."*

—UNKNOWN[2]

Carl Sagan, in his famous book on the cosmos, talked of the "billions upon billions" of stars that exist in our universe. The number of stars in the observable universe has been estimated to be on the order of three hundred sextillion: 300,000,000,000,000,000,000,000—a number that sounds fictitious, like a zillion or a gazillion. If you are anything like me, it makes you feel small and insignificant, and in awe of nature.

The amazing thing is that this is a *tiny* number compared to the number of different networks of friendships that could potentially exist among a small community—say a classroom, a club, a team, or the workers at a small company. Impossible, you say? How can this be so?

Consider a community of 30 people—for instance, all the parents of children in a class at school. Pick any one of our 30 parents—say Sara. Let us consider her friends within this community to be the people with whom she regularly talks or could depend upon to help her out. There are 29 other people with whom Sara could be friends. The second person—say Mark—not counting his potential friendship with Sara, could be friends with any of the 28 others. If you keep adding these up, the number of pairs of people in our small community who could be friends with each other is $29 + 28 + 27 + \ldots + 1 = 435$. Although that does not sound like too many possible friendships, it translates into a huge number of possible networks.

For example, if our community were completely dysfunctional, nobody would be friends with anyone else; we would have an "empty" network, devoid of relationships. So, all 435 possible friendships would be absent. If our community were completely harmonious, we would see the opposite extreme—a "complete" network in which every person would be friends with every other. There are many networks between these extremes. Maybe the first pair of people are friends with each other, but the second pair are not; then maybe the third and fourth pairs are friends, but not the fifth and sixth and so on. To find the total number of networks of friendships, we note that each possible friendship could either be switched "on" or "off," and so there are 2 possibilities for each friendship. Thus, the number of possible networks is $2 \times 2 \times \cdots \times 2$, with 435 entries. Doubling a number 435 times results in a 1 followed by 131 zeros—the sextillions previously mentioned have just 23 zeros.[3] So: sextillions of sextillions of sextillions of . . . networks—many times the number of stars in the universe, in fact, many orders of magnitude larger than the estimated number of atoms in the universe![4]

Even with just 30 people, there are far too many networks to label in any systemic way. In classifying animals, when someone says "zebra" or "panda" or "crocodile" or "mosquito" we know what they are talking about. Except for a few special classes, we really cannot do that with networks. This does not mean that we should throw up our hands and say that social structure is too complicated to understand.

There are also characteristics that allow us to classify and distinguish animals: Do they have a spine? How many legs do they have? Are they herbivores, carnivores, or omnivores? Do they have live births? How large are the adults? What type of skin do they have?

Can they fly? Do they live underwater? . . . When classifying networks we can identify critical characteristics too. For example, we can distinguish networks by the fraction of relationships that are present, whether those relationships are evenly distributed among the people involved, and whether we see certain segregation patterns. Moreover, these patterns will enable us to understand such issues as economic inequality, social immobility, political polarization, and even financial contagions.

Describing networks for our purpose of understanding human behavior is manageable for several reasons. First, a few primary features of networks yield enormous insight into why humans behave the way they do. Second, these features are simple, intuitive, and quantifiable. Third, human activity exhibits regularities that lead to networks with special features: it is easy to distinguish a network formed by humans from one in which the links are just formed randomly without any dependence on the other links around them or which nodes they connect.

As an example, consider the two networks in Figure 1.1. The network in panel (a) is a network of close friendships between high school students (details about this network appear in Chapter 5). The network in panel (b) has the same number of nodes and connections, but with the connections placed completely randomly by a computer.

So what is so different about the two networks? You can see a couple of things just by looking carefully. One is a sad fact of high school: there are more than a dozen students who have no close friends, while the random network has all nodes connected. The second more striking and general feature of the human network is that it is highly segregated. The students in the top part of the network are very rarely friends with the students in the bottom part of the network. The random network has links going in all directions.

The split in the network gets much easier to see, and more telling, when I add the races of the students in the high school, as in Figure 1.2.

Such divisions are one key feature of human networks, among several, that figure prominently in what follows. Why we form networks that have such features has some obvious explanations as well as some subtle ones, as we shall see. Ultimately, we care about our networks and their features because of their impact, and so by the end of this book you should know, for instance, why having divisions

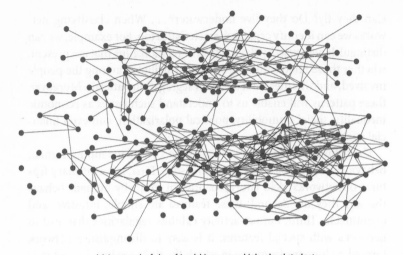

(a) A network of close friendships among high school students.

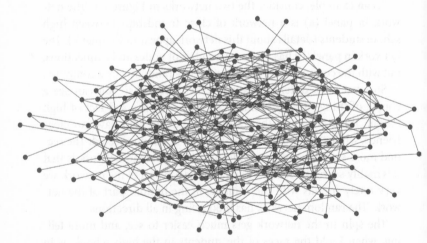

(b) A random network with the same number of links.

**Figure 1.1: A human network and a random network.**

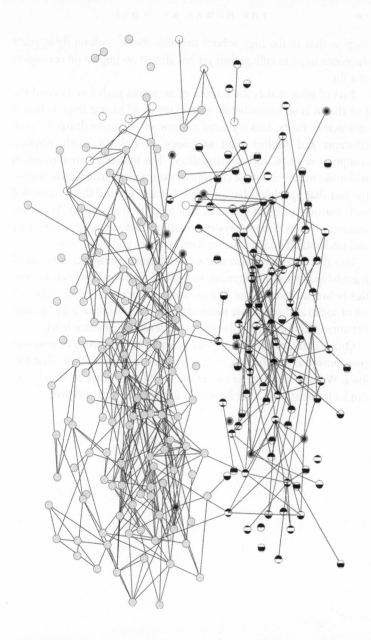

**Figure 1.2: The High School Network Coded by Race.** The nodes with bold stripes are self-identified as being "Black," the nodes with gray fill are "white," and the few remaining nodes are either "Hispanic" (center dot fill) or "Other/Unknown" (blank).[5]

such as that in the high school network above profoundly impacts decisions to go to college, but yet has almost no impact on contagion of a flu.

Part of what makes the science of networks such fun, beyond the fact that it is so immediately important in all of our lives, is that it cuts across fields: making sense of human networks draws on core concepts and studies from sociology, economics, math, physics, computer science, and anthropology.[6] For instance, our discussion will make heavy use of the concept of externalities from economics— the fact that people's behavior impacts those around them—coupled with various forms of feedback that amplify that impact. These are features of many complex systems: settings that are simple to describe and understand and yet rich in their features and behaviors.

Our discussion will also take us well beyond networks of personal friendships and acquaintances, to include relationships such as treaties between countries as well as contracts between banks. The full set of social and economic networks that we consider are all "human networks," as they all involve human interaction at some level.[7]

Our starting point is how your position in a network determines your power and influence, as this matters in almost all of what follows. We will make sense of the many different ways in which you can be influential and see how each depends on your network.

## 2 · POWER AND INFLUENCE:
## CENTRAL POSITIONS IN NETWORKS

*"Sometimes, idealistic people are put off by the whole business of networking as something tainted by flattery and the pursuit of selfish advantage. But virtue in obscurity is rewarded only in Heaven. To succeed in this world you have to be known to people."*

—SONIA SOTOMAYOR, *MY BELOVED WORLD*

Mahatma Gandhi mobilized tens of thousands of people to participate in the Salt March in 1930 to protest British rule. It was a walk of more than two hundred miles from Gandhi's base to the town of Dandi near the sea, where salt was produced from seawater. The narrow purpose of the march was to protest a salt tax. In such a hot climate, salt is essential and is consumed in large quantities, and high salt taxes were particularly symbolic of the hardships imposed on India by the British colonialists. More generally, the Salt March put in motion the acts of civil disobedience that would eventually end British rule.

If you see a parallel to earlier protests of British taxes on its colonies, you are not alone. The Boston Tea Party that protested British taxes more than a century before was not lost on Gandhi. In fact, he stated, "Even as America won its independence through suffering, valour and sacrifice, so shall India, in God's good time achieve her freedom by suffering, sacrifice and non-violence." It is said that after the Salt March, at Gandhi's meeting in London with Lord Irwin (the Viceroy of India), when asked if he wanted sugar or cream for his tea, Gandhi replied that no, he preferred salt "to remind us of the famous Boston tea party."[1]

The Salt March offered just a glimpse of what Gandhi would later accomplish, and his act of illegally producing salt in April of 1930 encouraged millions to follow in civil disobedience. Martin Luther

King Jr. mentions being moved when he first read of Gandhi's march to the sea, and it is easy to see how it inspired King's approach to the civil rights movement and organized marches.

These are examples in which an individual had the ability to, directly and indirectly, encourage millions of people to act. That reach was essential in Gandhi's and King's eventual success in changing the world. Judging power and influence by how many people a person can mobilize or impact is a natural starting point as it captures a person's reach.

Networks help us to identify and measure this sort of reach. A first measure of reach is simply counting how many people one knows or can count as a friend or colleague. In today's world we might also ask how many followers one has on social media. As we shall see, how many friends and followers a person has matters in subtle ways in driving a population's perceptions and social norms.

However, having many direct friends or connections is just one way in which a person can be influential, and much of this chapter will be devoted to understanding other network sources of power. Neither Gandhi nor King directly knew more than a small fraction of, nor could they personally contact, everyone they mobilized. They had key allies and friends, and also reached many through the publicity that their acts created. The Salt March began with a contingent of dedicated followers and swelled as it progressed and its publicity grew.

A person can have few friends or contacts and still be very influential if those few friends and contacts are themselves highly influential. This sort of indirect reach is often where power resides, and we can see this sort of influence very clearly via network concepts. Gaining influence via influential friends becomes an iterative and somewhat circular notion, but one that turns out to be quite understandable in a network context, with many implications. Iterative, network-based measures of power and influence will help us understand how to best seed a diffusion, as well as what it was that made Google an innovative search engine.

When it comes to measuring power, this will not be the end of our story. Another way in which people can be important, and one that is particularly evident when considering networks, is being a key connector or coordinator. A person can be a bridge or intermediary between people who don't know each other directly—enabling that person to broker favors and consolidate power by being uniquely positioned to coordinate the actions of others. This sort of power is seen in stories like *The Godfather,* and is evident in networks that explain the rise of the Medici in medieval Florence.

Understanding how networks embody power and influence will be useful when we later discuss things like financial contagions, inequality, and polarization. We will start with a look at direct influence.

## *Popularity: Degree Centrality*

Although he did not mobilize people to march like Gandhi, Michael Jordan did mobilize people to buy shoes. His ability to influence huge numbers of people was unparalleled. It is not by accident that, just during his sports career, Michael Jordan was paid more than half a billion dollars by companies wanting to advertise their products.[2] He earned just over $90 million in salary from actually playing basketball. By that metric, his value in marketing was (and remains) much larger than his direct value as an athlete and entertainer. Michael Jordan's incredible visibility enabled him to directly influence the decisions of millions of people around the globe.[3]

In network parlance, how many connections or links (relationships) that a person has in some network is called that person's "degree." The associated measure of how central a person is within the network is known as "degree centrality." If someone has 200 friends and someone else has 100 friends, then the first person is twice as central according to degree centrality. Such a count is instinctual and an obvious first method of measuring influence.[4]

And it is not just at the scale of a Gandhi, King, or Jordan that the number of people whom someone can reach matters. You are constantly being influenced by your friends and acquaintances. The people with the highest degree in any community, however small, have a disproportionate presence and influence.

What I mean by disproportionate presence refers to an important phenomenon known as the "friendship paradox," which was pointed out by the sociologist Scott Feld in 1991.[5]

Have you ever had the impression that other people have many more friends than you do? If you have, you are not alone. Our friends have more friends on average than a typical person in the population. This is the friendship paradox.

In Figure 2.1, we see the friendship paradox in a network of friendships in a high school from a classic study by James Coleman.[6] There are fourteen girls pictured. For nine of them, their friends have on average more friends than they do. Two have the same number of friends as their friends do on average, while only three of the girls are more popular than their friends on average.[7]

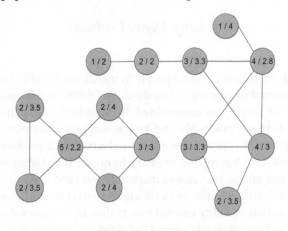

Figure 2.1: The friendship paradox. Data from James Coleman's 1961 study of high school friendships. Each node (circle) is a girl and a link indicates a mutual friendship between two girls. The paradox is that most of the girls are less popular than their friends. The first number listed for each girl is how many friends the girl has and the second number is the average number of friends that the girl's friends have. For instance, the girl in the lower left-hand corner has 2 friends, and those friends have 2 and 5 friends, for an average of 3.5. So the 2 / 3.5 represents that she is less popular than her friends on average. This is true for 9 out of 14 of the girls, while only 3 are more popular than their friends, and 2 are equal in popularity to their friends.

The friendship paradox is easy to understand. The most popular people appear on many other people's friendship lists, while the people with very few friends appear on relatively few people's lists. The people with many friends are *overrepresented* on people's lists of friends relative to their share in the population, while the people with very few friends are underrepresented. Someone with ten friends is counted as a friend by twice as many people as another person who has just five friends.

In a mathematical sense, the paradox is not very deep—but paradoxes rarely are. Nonetheless it has implications for almost all of our interactions. Anyone who has been a parent, or a child for that matter, is familiar with statements like "everybody else at school has a . . ." or "everybody else at school is allowed to . . ." Although these sorts of statements are usually false, they often reflect what we perceive. The most popular students can be greatly overrepresented among the children's friends, and so if the most popular students are all following some fad, then children end up thinking that everyone else is. Popular people disproportionately set perceptions and determine norms of behavior.

To see the implications of the friendship paradox most starkly, let us consider a simple example, and then look at some data that corroborate the example.

Consider a class of students who are influenced by their friends.[8] These students, deep down, are conformists. They are faced with a simple choice: do they wear solid or plaid clothes? They each have a preference for solid or plaid and on the first day of school they follow that preference, as pictured in Figure 2.2.

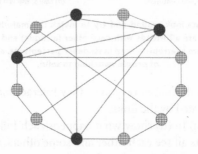

**Figure 2.2: The first day of school: The four most popular students have a preference for solids; the eight others prefer plaids.**

Being true conformists, the students would like to do what the majority of others are doing, and only follow their own preference if there were equal numbers of others in each style. As pictured in Figure 2.2, four students prefer solids and eight prefer plaids. Thus, two thirds of the students prefer plaids, and if they could all see the whole group's preferences, then they would all wear plaids the next day. However, note that it is the four most popular students, perhaps the boldest students, who prefer solids.

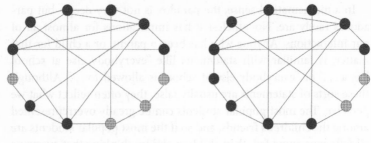

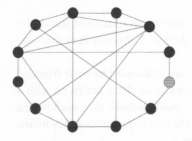
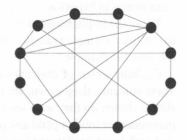

(a) Day 2, Four switch to match the popular students.

(b) Day 3, A cascade, as more switch.

(c) Day 4, The cascade continues.

(d) Day 5, the last holdout conforms.

**Figure 2.3: Students look around and try to match the majority of their friends. The most popular are all friends with each other (a clique) and all stay with solid. Popular students are overrepresented in students' perceptions, and begin a cascade of people switching to solid.**

The students don't see everyone—they interact mainly with their friends as indicated by the links.

Figures 2.3 (a) to (d) show what happens each following day. The popular students all see each other and some others, and they all see a majority wearing solids and so they continue to wear solids. Some other students see mostly popular students, and so they switch to wearing solids. As we see in Figure 2.3 panel (a) the popular students all stay with solids and four more students switch to solids, and by the second day we have eight of the students wearing solids. Things quickly unravel from there, as we see in panels (b) to (d). Each day more of the students who are still wearing plaids see a majority of their friends wearing solids and they switch to solids. By the fifth day every student in the class ends up wearing solids, despite the fact that a majority of them started with a preference for plaids.

We can see the friendship paradox's role in this cascade of fashion by examining Figure 2.4, which shows how the students incorrectly perceive the population preferences, based on what they see among their friends on the first day. The most popular students are overrepresented in people's friendships and so three quarters of the students perceive that solids are in the majority even though two thirds of the students prefer plaid.

There are two aspects that you might notice about the structure of this example. One is that the most popular students all have the same preferences: all like solids. This helps in speeding coordination, and to their preferred fashion. This matters, and there are reasons for why the most popular students will be similar to each other, as we shall soon explore. The second is that the popular students form a clique—they are all friends with each other. This reinforces their behaviors and maintains their norm of solids, which then eventually takes over the rest of the population. It makes the example work cleanly, but the idea that the most popular people are disproportionately influential still holds without this. Indeed, fashion designers have long understood the importance of having celebrities wear their new and different designs on the Oscars red carpet.

The impact of popularity and the friendship paradox is perhaps at its purest in settings of peer influence, such as students' perceptions of others at school. A long series of studies have found that students tend to overestimate the fraction of their peers who smoke,

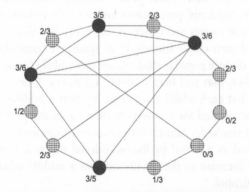

**Figure 2.4: The friendship paradox at work. The fractions next to the students are their perceptions of the preferences for solids over plaids, based on what they see among their friends. Most of them mistakenly perceive a majority preference for solids, with only the few students in the lower right initially perceiving a majority for plaids. Even those students will quickly see a majority wearing solids.**

consume alcohol, and use drugs, as well as the frequency with which they do so, and often by substantial margins. For instance, a large study covering one hundred U.S. college campuses found that students systematically overestimate consumption of eleven different substances, including cigarettes, alcohol, and marijuana.[9] In particular, a further study focusing on alcohol consumption compared students' self-reported drinking behavior—how many drinks they had the last time they partied or socialized, with their perceptions of how many the typical student at their school had the last time she or he partied or socialized. Out of the more than 72,000 students at the 130 colleges in the study, the median student answered 4 drinks—a fact that seems alarming, especially given that a quarter of the students answered 5 drinks or more. But what is surprising, given how high these numbers are, is that more than 70 percent of the students still managed to overestimate the alcohol consumption of the typical student at their own school by a drink or more.[10]

To explain these misperceptions, we don't have to dig deeply into the psychology of the students. The friendship paradox provides an easy insight. When students are attending parties or social events, they are interacting disproportionately with the people who attend the most parties—so students' perceptions of alcohol consumption ends up overrepresenting people who attend many parties. This is a version of the friendship paradox—the people who students see at the parties are more likely to attend more parties than the average student. Students' perceptions are not only influenced by their experiences at parties and other social events, but also by what they know about their closer friends. Here again, the friendship paradox is at work. If more popular students are more likely to smoke or consume alcohol, that will bias students' estimates. Indeed, one study estimated that each additional friendship that a middle school student had accounted for a 5 percent increase in the probability that the student smoked.[11] Similar estimates have been found for alcohol: being named as a friend by five additional others accounted for a 30 percent increase in the likelihood that a middle school student had tried alcohol.[12]

There are several effects that push students who socialize the most to be higher consumers of alcohol and cigarettes. One is that such consumption by teenagers is a social activity. People who spend more time socializing with others thus have more reason to consume alcohol. There is also a reverse effect: students with a higher propensity to consume alcohol would tend to seek out opportunities to consume it and others with whom to share it.[13] On top of these effects is that students who have less parental supervision have more time to hang out with other students, and more opportunities to try alcohol, cigarettes, and drugs. Finally, social activities by their very nature experience feedback. Viewing one's peers drinking encourages drinking. That increased level of drinking further increases the drinking of peers, and so this continues to cycle in a feedback loop.[14]

So, given that students' estimates of peer behavior are based at least in part, if not largely, on their personal observations, the friendship paradox and the fact that the most socially active students often take more extreme behaviors leads us to expect students to systematically overestimate peers' behaviors. More generally, given that many behaviors are influenced by perceived norms, we end up with behaviors driven disproportionately by those who socialize the most, and the resulting norms are more extreme than if our perceptions were not network-based.

The friendship paradox is enhanced by social media, where the magnitude of the effect can be staggering. For example, a study of Twitter behavior[15] found that more than 98 percent of users had fewer followers than the people whom they followed: typically a user's "friends" had more than ten times as many followers as the user. Those more popular users are more active, and despite their small numbers they play an important role in viral content. Given the increased use of social media, especially by adolescents, the potential for biased perceptions in favor of a tiny proportion of the most popular users becomes overwhelming, especially when one factors in that the most popular social media users may have very different behaviors, as we see from the relationship between students' popularity and earlier and heavier use of alcohol and cigarettes. Partying is also by its very nature a social activity, which can further amplify the effects of social media as pictures and stories of alcohol and drug consumption are shared. In contrast, behaviors such as studying tend to be more solitary events and information about them is less likely to be shared. It is thus natural for a teen to overestimate the amount of drugs and alcohol consumed by his or her peers and to underestimate the time spent studying by those same peers.

The bias that accompanies the friendship paradox, whether or not we realize it, applies well beyond "friendships." The friendship bias is an example of "selection bias": our observations are often from biased samples depending on how that sample was picked. We disproportionately fly on the most heavily booked flights, eat at the most popular restaurants, drive on the busiest roads and at the busiest times, go to parks and attractions at the most crowded times, and attend the most crowded concerts and movies. These experiences bias our perceptions as well as our perceived social norms, usually without our understanding those effects. As Shane Frederick (2012) states in a study about our tendency to overestimate other people's willingness to pay for things: "Customers in the queue at Starbucks are more visible than those hidden away in their offices unwilling to spend $4 on coffee."[16]

## Comparisons, Comparisons

*"If you torture the data enough, nature will always confess."*

—RONALD COASE, *HOW SHOULD ECONOMISTS CHOOSE?*[17]

*"I want to be perceived as a guy who played his best in all facets, not just scoring."*

—MICHAEL JORDAN, 2003 NBA ALL-STAR GAME

Who was the best basketball player of all time, Wilt Chamberlain or Michael Jordan? Maybe you would like to make a case for LeBron James. Having grown up in Chicagoland, I have my own answer to such questions, but the comparison is really between great athletes who had very different styles and roles in the game.

There are many different statistics that can be used to summarize their careers. For instance, Jordan and Chamberlain are amazingly similar on several dimensions: they each averaged 30.1 points per regular season game during their careers, both had just over 30,000 points in regular season games (32,292 for Jordan and

31,419 for Chamberlain), and each amassed several Most Valuable Player awards (5 for Jordan and 4 for Chamberlain). However, there are other dimensions on which they differed: Michael Jordan led his team to more NBA championship wins (6 to Wilt's 2), but Wilt Chamberlain amassed dizzying numbers of rebounds per game (22.9 to Michael's 6.2).

There are other dimensions on which other players stand out. Steph Curry's record three-point totals are far beyond anything seen before. Kareem Abdul-Jabbar's longevity at a high level is unparalleled. Kareem played for 20 years, amassing more than 40,000 points in total, and played in 19 All-Star Games, after having dominated basketball at the college level as nobody had before. LeBron James's all-around dominance has been evident since he appeared on *Sports Illustrated*'s cover as a junior in high school. But if we really want to measure all-around contributions then we should consider the triple-double—having at least 10 points, 10 rebounds, and 10 assists—all three statistics in double figures. Then one has to remember Oscar Robertson, who *averaged* a triple-double for a whole season (a feat only recently matched by Russell Westbrook), and had so many triple-double games that nobody else even comes close, not even Magic Johnson.

The point here is not really to have a "da Bears, da Bulls"[18] argument about basketball prowess, but to emphasize several things: statistics capture useful information in a succinct manner, different statistics encapsulate different things, and even a long list of statistics can fail to capture all of the nuances of the things that they describe.

Our lives would be simpler if measuring something could always be boiled down to a single statistic. But part of what makes our lives so interesting is that such unidimensional rankings are generally impossible for many of the things that are most important to understand: lists of rankings end up being both controversial and intriguing. How does one compare the musical innovations of Haydn, Strauss, and Stravinsky; or the contributions to human rights of Eleanor Roosevelt, Harriet Beecher Stowe, and Harriet Tubman? Is Lionel Messi or Diego Maradona the more impressive soccer player? Can one possibly compare the art of Pablo Picasso to that of Leonardo da Vinci? Or, is it easier to compare the paintings of Pablo Picasso to those of Henri Matisse, not only because Picasso and Matisse were contemporaries, but because they were rivals? Many might argue that such comparisons are hopeless and meaningless. However, they force us to think carefully about the various dimensions on which these people made contributions and why those contributions were game-changing.[19] When one looks at different basketball statistics one sees different players stand out, with each amazing in his or her own way. Similarly, when looking at different statistics that characterize people's network positions, different people stand out as being most "central." Some people end up being very central according to some but not other measures, and which network statistic(s) are most appropriate depends on the context, just as whether you would rather add a top scorer or a top defender to your basketball team would depend on the circumstances.

We have already seen that one measure of centrality—degree centrality—helps us understand why the highest-degree people in a network end up having disproportionate influence. This is a first "network effect." As the most basic and obvious measure of network centrality, degree centrality is akin to average points per game in the basketball example. However, to complete the analogy, different people can have different strengths in terms of their positions in networks—so that who is most "central" will vary with the way in which we ask the question, just as Wilt was a dominant rebounder while Michael drove his team to championships and Steph Curry stretched defenses in new ways. Comparing nodes (e.g., people) in a network based on their degree centrality can completely miss some of the most essential aspects of power and influence. So let's see some other concepts.

## It's Who You Know—Locating the Needles in the Haystack

*"Networking is rubbish; have friends instead."*

—STEVE WINWOOD

Google might not even exist except for the serendipitous assignment of Sergey Brin to show Larry Page around the Stanford University campus in 1995, when Larry was considering Stanford for his doctoral studies. Sergey's family had emigrated to the U.S. from Russia in the late 1970s. Long fascinated by mathematics and computer programming, Sergey had come to Stanford for its computer science program. Larry Page shared a similar fascination with computers, and recalls a childhood of "poring over books and magazines, or taking things apart at home to figure out how they worked." Although their strong personalities clashed at times, their common interests and intellects led them to a fast friendship. Most important for us, they shared a growing curiosity about the structure of the World Wide Web.

By 1996 Sergey and Larry were working together on the design of a search engine for the Web. They began using Larry's dorm room to house a set of computers cobbled together out of the parts that they could find, and Sergey's room for an office where they developed their ideas and programs. In a paper that they wrote together as students, Sergey and Larry describe how rapidly the Web was expanding in the late 1990s and how search engines were not really up to the task. One of the first search engines, the World Wide Web Worm of 1994, indexed just over 100,000 pages. By 1997, another search engine, AltaVista, was claiming to have tens of millions of queries per day, and the Web already had hundreds of millions of pages to search and index. The sheer volume of pages to index was making it impossible to find what the user wanted. To quote Brin and Page, "as of November 1997, only one of the top four commercial search engines finds itself (returns its own search page in the top ten results in response to a query of its name)."

So how does one locate the right needles in such a giant haystack? There are some obvious ideas as to how to identify the Web pages that a user might want to see when they type in some keyword. But huge numbers of pages contain the same keywords. Having the keyword appear frequently on some page does not come close to guaranteeing that it is what most users are looking for. Perhaps tracking past traffic and looking deeper into the content of various pages might help. Many variations on this theme were being tried but nothing seemed to work adequately. It was easy to begin to think that the Web was just becoming too large, and indexing and navigating it in any sensible way was destined to be an overwhelming task.

Brin and Page's breakthrough was born from their interest in the network structure of the Web: it holds a lot of useful information, as the structure is not an accident. Web pages link to other Web pages that they see as being important. So how did Brin and Page understand and use that information? Brin and Page's key insight was that a useful way to identify a page that a searcher might be most interested in was to look at *which* other Web pages have links pointing to that Web page. If other important Web pages point to a page, then that suggests that it is an important page. One does not judge a page simply by how many pages link to it, but by whether it is linked to by well-connected pages. In many settings it is more important to have "well-connected" friends than just to have many friends.

This sort of definition is circular: a page is "important" because it is linked to by other "important" pages, which are in turn "important" because they are linked to by other "important" pages. Despite the circularity, it turns out to have a beautiful solution, and one that is extremely helpful in network settings.

Suppose we want to spread a rumor or some information that we think will be relayed via word of mouth. To see why a straight measure of popularity falls short, consider the network in Figure 2.5. It is clear from just looking that the positions of Nanci and Warren are quite different from each other, even though they each have two friends. They differ in terms of how well-connected their friends are, and relatedly, how well-positioned they are in the network. Warren's friends have only two friends each, while Nanci's friends have seven and six friends. So, while Warren and Nanci score equally well in terms of their "degree" (number of friends), Nanci's friends have higher degrees than Warren's.

We could stop here: instead of just counting friends, we could count how many additional friends each of those friends brings—so we could track friends of friends: which we can call "second-degree

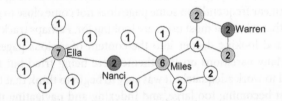

Figure 2.5: Two people, Nanci and Warren, both have degree 2. However, they differ in how connected their friends are and in their overall positions in the network.

friends." Looking beyond direct friends to count friends of friends would be a good start, and Nanci already becomes clearly better situated to spread information than Warren. But why stop iterating here? Why not consider "third-degree friends"? Now Nanci's friendship with Ella is not so fruitful in terms of third-degree friends, but her friendship with Miles leads to even more connections. By the time we have gone out three steps from Nanci we have reached everyone except Warren. Going out three steps from Warren we only reach five other people, while from Nanci we have reached sixteen. This makes Nanci a much better candidate for spreading information than Warren, even though they both have the same degree.

How does one capture this in a large network, as we could go on forever? There are various ways of doing so, but let me describe the crux of the idea. Let us start by just adding up first-degree (direct) friends. So, as we see in Figure 2.5, Nanci and Warren each get a value of 2 since they each have two friends. Next, let's add in second-degree friends. But should we count these as highly as first-degree friends? For example, if we think about spreading information starting with Nanci, it is more likely that information gets from Nanci to Miles, than to a friend of Miles—as it first has to pass from Nanci to Miles and then also has to be spread further from Miles. It might be much less likely to make it two steps than just one step, for instance half as likely. So, for now, let's weight a friend of a friend half as much as we value a friend. Nanci has eleven second-degree friends, so she gets a score of 11/2 for her friends-of-friends. Warren has only one second-degree friend and so he gets 1/2. Thus, Nanci has a score of 7.5 so far, counting first- and second-degree friends, while Warren's score is now only 2.5. As we move out to third-degree friends, Nanci has three, while Warren has two. Again let's weight those by another factor of a half, so we will give each of them a score of 1/4. So Nanci adds 3/4 to her score and Warren adds 2/4 to his score, and so Nanci is up to 8.25 and Warren is up to 3. Iterating in this manner, we can quantify how much more reach Nanci has in the network than Warren.

The relative comparison between Nanci and Warren also turns out to be the solution to another question. Let's define each person's centrality as being proportional to the sum of their friends' centralities. This is similar to the calculation we just did. By doing this, Nanci gets some fraction of Ella's and Miles's scores, which come from adding up some fraction of their friends' scores, and so forth. This iteration is similar, because Ella's and Miles's scores are coming from their friends, which are Nanci's second-degree friends, and those scores come from their friends, which are Nanci's third-degree friends, and so on.[20]

Luckily, this type of system of equations, in which each person's centrality is proportional to the sum of her friends' centralities, is a quite natural and manageable math problem. It developed through a series of contributions from a who's who list of mathematicians from the eighteenth through twentieth centuries: Euler, Lagrange, Cauchy, Fourier, Laplace, Weierstrass, Schwarz, Poincaré, von Mises, and Hilbert. Hilbert named the solutions to such problems "eigenvectors" (pronounced "eye"), the common modern name. Not surprisingly, eigenvectors pop up in all sorts of applications from quantum mechanics (Schrödinger's equation) to the definitions of the "eigenfaces" that comprise the basic building blocks used in facial recognition patterns. When solving for the eigenvector in our example, we find that Nanci's score is about 3 times that of Warren, as we see in Figure 2.6.[21]

The Brin and Page innovation was to rank Web pages by what they termed PageRank—which relates to our discussion above and to an eigenvector calculation. Although Brin and Page's problem was not spreading a rumor through a network, it was based on another closely related iterative problem called the "random surfer problem." A user starts at some page and then randomly follows a link from that page to another page, with each link getting equal probability. The user then repeats this, randomly surfing the Web in this fashion.[22] Over time, if we calculate the relative fraction of times that the user lands on each page, it is an eigenvector calculation. In this case, the weights that are being used at each step are proportional to the number of links embedded in each page.

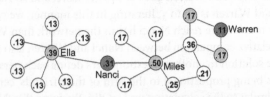

**Figure 2.6: Eigenvector centralities of each node (person). Nanci outscores Warren by almost a factor of 3, even though they both have the same number of connections. Miles ranks highest, even though Ella has the highest degree centrality.**

There were two challenges that Brin and Page faced. The conceptual challenge of finding the most relevant pages was addressed by not just ranking pages by popularity, but by calculating how "'well-connected" the pages were in this iterative, eigenvector sense. The more practical challenge was implementing this on the huge scale of the Web, which involves crawling the Web and indexing pages, storing data about the content and links of each page, and then making such iterative calculations about network position. It is one thing to calculate such things for Nanci and Warren in our small network above, but it is another to approximate this for billions of pages, especially when they are constantly evolving in their content and links.

Brin and Page developed an algorithm based on these sorts of calculations, and well-suited for huge networks, and called it BackRub, which they started running on Stanford servers. The name BackRub comes from looking at backlinks—the links that lead one to a page. BackRub quickly outgrew the student accounts that Brin and Page had on the Stanford servers, and by 1997 they had moved the search engine and renamed it Google—a variation on googol, which is the number corresponding to a 1 followed by 100 zeros, referring to the vast size of the Web that their algorithm managed to conquer. For anyone who struggled through the early days of searching the Internet, the ability of Google to find useful pages was incredible. There were many competing search engines, and typically one would try several search engines in an often futile attempt to find a Web page that one really needed. By 1998, *PC Magazine* reported that Google "has an uncanny knack for returning extremely relevant results" and ranked it in its top hundred Web sites.[23] The rest is history.[24]

## *The Diffusion of Microfinance*

Although Google history suggests that an eigenvector-centrality-based algorithm outperformed the alternatives, search engine algorithms are complicated and other differences in their algorithm might have also explained Google's success. It would be nice to see more definitive evidence that one's friends' positions matters. Also, BackRub was identifying pages by how easy they are to reach, while in many situations we are interested in how influential someone is in terms of reaching others.

This was on my mind when I was visiting MIT in 2006 and talking with Abhijit Banerjee, a friend and professor there, and discussing how it would be wonderful to really test such differences in action. As luck would have it, Abhijit was precisely the right person for me to be talking to (as he often is). It turns out that Esther Duflo, another MIT professor, was in touch via her sister Annie with a bank in southern India, called BSS (Baratha Swamukti Samsthe), that was planning to roll out a new microfinance program via word of mouth. (You can see the network at play even in how this research project got off the ground.) The word-of-mouth program ended up offering us a perfect opportunity to see how network structure mattered in the spreading of information, and ended up allowing us to test which centrality measure would best predict a person's ability to spread information. Abhijit, Esther, and I, together with Arun Chandrasekhar, who was then a graduate student at MIT (and coincidentally whose family was from Karnataka, the region in question), began what turned out to be a long-term study.

The pioneer of the microfinance revolution was Muhammad Yunus. He founded the Grameen Bank in Bangladesh in the 1970s and began making widespread loans of very small amounts in the 1980s. Yunus and the Grameen Bank were recognized for their innovation with a Nobel Peace Prize in 2006. The innovation was simple but clever. Many loans throughout the world involve a house or a car as collateral, or are advances on paychecks to people with an employment record, or are a loan via a credit card for people with proven credit history, backed by aggressive collection agencies that go after defaulters. Microfinance loans are aimed at extremely poor people with variable employment, and little to no collateral, and in settings where trying to collect would be prohibitively expensive. So, what was the innovation?

The innovation was that the loans were based on joint liability— holding several people responsible if someone failed to repay their loan. If someone defaults on their loan, their friends also feel the consequences. Now there are many variations of such microfinance loans, but a typical system is the one followed by the bank in our story, BSS, and illustrates the idea. BSS's loans were offered exclusively to women between the ages of eighteen and fifty-seven, with a limit of one loan per household. Women were formed into groups of five who were held jointly liable for the loans: if one of the women defaulted on her own loan, then the entire group was called into default on their loans. Default then denies a borrower access to future loans—or at least makes it more difficult for the defaulter to borrow again. In some cases, this operates at an even wider level in which joint liability extends across groups, so that too many defaults would cut a village off from a lender entirely. Holding people jointly liable for repayments leads to reputational and social pressures on people to not let their fellow villagers down by defaulting, and also means that group members have incentives to help each other out and step in if someone is unable to pay.

Also, repayment of one loan typically enables the borrower to take out subsequent loans, which then increase in size. The promise of larger future loans based on current repayments—essentially allowing these people to build a credit history one step at a time—was another big incentive to repay. In addition, participants often receive some basic financial training that encourages some savings and teaches them to track income, plan, and how to keep a simple book tracking payments. While this training may seem rudimentary, it can be empowering to the villagers.[25] On a visit to one of the villages, a woman being interviewed about her finances gave me an illuminating lecture on how she had been increasing the size of her loans, was maintaining an accounting system tracking money in and out of the household, had built better-diversified groups involving both Muslims and Hindus, and had put together several loans to buy a used truck and start a business.

Although there were some late payments, defaults on loans issued by BSS in these villages in the years of our study were almost nonexistent.[26]

Another important aspect of microfinance is that the restriction of the loans to women impacts the dynamics of a household. Even though some, if not much, of the money ends up in the control of males within the households in such villages, the fact that the loans can enter only via a woman in a household can give the women some say in how the money ends up being invested or spent.[27]

BSS's spread of microfinance illustrates the importance of network centrality, and the difference between degree and eigenvector centrality.

The bank BSS in our study was faced with the question of how to disseminate the news about the availability of microfinance to potential borrowers in the seventy-five villages in Karnataka that it was planning to enter. The volatile and caste-based politics of the villages, coupled with corruption, meant that the bank did not wish to rely on local village governments to spread information. Although some villagers can be reached via cell phones, they are so bombarded with spam texts that advertising via phones was also not viable. Posting flyers and even driving around with a loudspeaker are other techniques for advertising; but again these are overused and primarily associated with political campaigning. So, for better or worse, the method that the bank settled on was to find a few "central" individuals and ask them to spread the word about the bank and availability of microfinance.

Without knowing the networks of friendships, how could the bank identify the most central villagers? Would it even matter? The bank guessed that the best-positioned villagers to spread information would be teachers, shopkeepers, and self-help group leaders.[28] Let us call these people "the initial seeds." Essentially the bank expected these initial seeds to be central—and the bank was thinking of degree centrality and had no concept of eigenvector centrality.

What was useful for our study was that in some villages the initial seeds did have high degree, while in other villages they happened to have low degree. For instance, in some villages a teacher had many contacts, but in another village the teacher did not. More important, there were also villages in which the initial seeds had high eigenvector centrality, but low degree centrality, and other villages with the reverse. Also, in some villages this technique of seeding information worked well, while in other very similar villages it failed miserably: the participation rate of eligible households in some villages was nearly half and in others it was less than one in ten. Thus, we could see which centrality measure best predicted the spread of information from the initial seeds. So, which measure of network centrality of the initial seeds explains the more than six-fold difference in eventual diffusion across villages?

In 2007—before BSS entered the villages—we surveyed the adult villagers and mapped out their networks. These small villages are especially well-suited for network analysis because most interactions are within the village and in person.[29]

Given our discussion of the importance of popular people in determining the perceptions of others and in setting trends, at first blush it makes sense that high-degree people should be good seeds for diffusing information about microfinance. This turned out not to be the case at all—there was no relationship between the initial seeds' degrees and the spread of microfinance in the villages.[30]

Was our discussion of the importance of popularity nonsense? Clearly not. As with basketball players, popularity can be important, but it is just one facet of a rich picture. Popular individuals play roles in creating perceptions of social norms and fads, and *directly* reaching people. However, we found in our study, the main issue in the microfinance villages was getting information out widely to a whole village rather than simply influencing perceptions. Even for people living in a remote village, by 2008 it was hard to be unaware of microfinance, just as most people in the developed world are aware of credit cards and know that it is useful to have one. This was not about creating a trend, or influencing villagers' perceptions of how many other villagers are taking out microfinance loans; it was a matter of making as many villagers as possible aware that loans were available.[31]

Indeed, spreading the news about microfinance was not simply about how many friends the initial seeds could reach, but also about how many friends-of-friends (second-degree friends) and third-degree friends, etc., that the initial seeds could reach.[32] The immediate friends of the initial seeds were generally just a small portion of the total populations. Despite the fact that degree of the initial seeds did not seem to matter much at all, there was significantly higher participation in villages in which BSS's initial seeds had higher eigenvector centrality than in villages in which the initial seeds had lower eigenvector centrality. Comparing a village whose initial seeds were at the bottom end of the eigenvector centrality to one whose initial seeds were at the top of the scale led to a tripling of the participation in microfinance, on average. Having information spread widely in the villages required getting it to flow well beyond the initial seeds' friends, to their friends of friends of friends. . . .

## Diffusion Centrality

This isn't quite the end of our microfinance story.

The interest in any topic eventually decays with time. Most news stories get the majority of the attention they will ever get within hours or days, and are quickly displaced by subsequent stories, not only in the media but also in people's discussions and further spreading of them. This means that eigenvector centrality might overdo the calculation of position in a network for spreading information. On the one hand, just looking at degree centrality misses the fact that news spreads beyond one step. On the other hand, eigenvector calculations look at an infinite process that keeps on cycling endlessly through the network. Reality is somewhere between these extremes.

With that in mind, for our microfinance analysis, we defined a new centrality measure to capture what happens in such real diffusion processes. People spread news, but stop communicating a particular topic after some number of iterations. For instance, a topic might be talked about for a couple of days, but then people lose interest. Our estimate for the spread of microfinance was that news tended to spread for roughly three iterations—not really traveling much beyond the friends-of-friends-of-friends.

Also, some topics inspire people to talk with everyone they know about it, and other topics inspire people less. In the case of microfinance, our estimate of the frequency with which one household would tell one of its friends at each iteration was 1/5. This would be like doing our Nanci and Warren calculation with a weight of 1/5 instead of 1/2, and just reaching to friends of friends of friends instead of iterating infinitely.[33] Nanci still outscores Warren, but not by as much.

Diffusion centrality spans between two extremes of degree centrality and eigenvector centrality. If one lets the number of iterations and the probability of information moving from one node to another become large enough, then diffusion centrality mimics eigenvector centrality, while with just one iteration or a tiny probability of transmission it becomes proportional to degree centrality. In between, it captures a limited reach of a person in their network, and adjusts for how topical and long-lived whatever being spread is.

Diffusion centrality turned out to be a much more accurate predictor of the spread of microfinance than even eigenvector centrality. The diffusion centralities of the initial seeds did several times better than their eigenvector centralities at explaining the differences in the spread of microfinance across villages.[34]

The ultimate moral of this story is that there are different ways to measure centrality, and some do better than others at predicting what will happen, depending on context.

So far, we have seen three distinct conceptual approaches to measuring a person's position in a network; with degree centrality identifying direct influence, eigenvector centrality capturing the power of one's friends, and diffusion centrality tracking the reach that someone has at spreading (or receiving) information with some limited time and interest. As our basketball analogy suggests, these are just a few among many ways to capture the importance of position in a network. Although we don't need to catalogue all of them, there is yet one more centrality measure that differs in important ways from what we have seen. A fascinating historical episode, the rise of the Medici, helps illustrate one of the most interesting centrality measures from a power perspective.

## *The Rise of the Medici: An Early Lesson in Networking*

*"The Medici created and destroyed me."*

—LEONARDO DA VINCI

*"Political questions are settled in [Cosimo's] house. The man he chooses holds office. . . . He it is who decides peace and war. . . . He is king in all but name."*

—PIUS II

Fourteen thirty-four was a pivotal year in Florence, ultimately shaping the patronage that fed the early Renaissance. Florence transitioned from being an oligarchy, ruled by various factions of wealthy and politically prominent families, such as the Albizzi and the Strozzi, to a society dominated by one family: the Medici. It is perhaps not coincidental that around this time Donatello's famously original and innovative life-sized bronze statue of David, commemorating a triumph of a hero over powerful odds, was commissioned by the Medici. What enabled Cosimo de' Medici, the family patriarch, to consolidate power?

The Medici, although elite, did not stand out politically or financially prior to the 1430s. For instance, the Strozzi had greater wealth and controlled more seats in the local legislature, and yet the Medici rose to eclipse them.

Prior to 1434, the Medici had struggled against the other oligarchs, including the Strozzi and the Albizzi, other wealthy and powerful banking families. That conflict came to a head when the Albizzi and Strozzi played prominent roles in exiling Cosimo de' Medici and other members of his family from Florence in 1433. The antagonism between the families was not only due to the current power struggles and economic burden of a lost war against the rival city of Lucca, but had deeper roots. The Medici, and cousin Salvestro de' Medici in particular, were key supporters of an uprising of wool makers and tavern owners against the more established guilds in the 1370s and 1380s. The uprising, known as the Ciompi (a term that translated roughly as "friend, let's get a drink"), was in opposition to heavy taxation and the attempts of the nobility to stop the lower classes from belonging to guilds and gaining political and economic power. Although the uprising ultimately failed, it resulted in lasting changes, and the Medici support for the uprising marked their name for decades to come. As the Medici banking empire continued to grow, their competition with the other oligarch families came to a head in 1433. At that time, the other oligarch families comprised most of the Signoria, the central and rotating political body ruling Florence that consisted of nine guild members who were called the Priori. In September of 1433, Bernardo Guadagni, a close ally of the Albizzi, became the Gonfaloniere of Justice—a job that rotated among the Priori. Rinaldo degli Albizzi, along with the Strozzi and other oligarch families who feared the Medici, helped convince Bernardo and the Signoria to banish Cosimo de' Medici and some of his family members from Florence. This took place after the Signoria held a hasty consultation with some assembled citizens under the watchful eye of the Albizzi's troops.

The exile of the Medici was not to last. The oligarchs opposing the Medici had greatly underestimated Cosimo's power. He and his allies managed to pull large amounts of capital out of Florence. Given the already heavy impact of the lost war with Lucca, this led to a severe financial crisis. In addition, via the Medici's many alliances with other families, Cosimo also influenced the selection of a new Signoria. This quickly turned the tables and Cosimo returned to Florence to a parade in the fall of 1434. A few days afterward, it was Rinaldo degli Albizzi who was banished—forever.

How was it that Cosimo had the power to rally many allies and coordinate retaliation and a change of government? Why weren't the Albizzi able to respond?

First, it almost goes without saying that Cosimo had to know what he was doing. Cosimo's consolidation and exercise of such power in a highly contentious environment took considerable foresight, skill, and intellect. That intellect and broad view of life are seen in Cosimo's interests in philosophy (he commissioned the first full translation of Plato's works), his widespread patronage of the arts (not just Donatello, but also Fra Angelico, Fra Filippo Lippi, Lorenzo Ghiberti, Michelozzo di Bartolomeo, and Filippo Brunelleschi), his sponsoring of the first "public" library in Florence, and his role as ambassador and in international politics. Cosimo was truly a Renaissance man, and his generosity, as well as his business, social, and political maneuvers, earned him the reverence of Niccolò Machiavelli writing almost a century later. Machiavelli writes, "Cosimo was one of the most prudent of men; of grave and courteous demeanor, extremely liberal and humane. He never attempted anything against parties, or against rulers, but was bountiful to all; and by the unwearied generosity of his disposition, made himself partisans of all ranks of the citizens." (Book IV of *History of Florence*.)

But second, and more important from our perspective, Cosimo's understanding of, and fortuitous position in, the mosaic of Florentine social and economic networks was essential. That position enabled him to build and control an early forerunner to a political party, while other important families of the time floundered in response.

The Medici's network involved two key sets of connections—business dealings and marriages. The business dealings centered around their bank, which consisted of franchises among extended family members. The Medici bank was a primary resource not only for the elite families of Florence, but also for many of the nonelites, and also catered to the papacy and many religious leaders throughout the region. Beyond basic banking and loans, the Medici were involved in a variety of partnerships, real estate dealings, and trade. These economic relations were complemented by a network of marriages to other elite families.

Marriages among elite families at that time were far from romantic affairs. Such a marriage might involve a son in his mid-thirties from one family marrying a young daughter in her teens from another family. The daughter served as a sort of social collateral, bonding her new family to her blood relations, and the son-in-law would often become an important business connection and political lieutenant for the family into which he married.[35]

These business and marriage ties embodied the allies and collateral that cemented relationships, enabling collaboration in an environment where political alliances and economic contracts were otherwise difficult to enforce and in which competition could be intense.

The unique position of the Medici is seen in the network of marriages and business dealings between some of the key elite families in Florence as pictured in Figure 2.7. Each node represents a family and the links connecting pairs of families represent the marriages that connected families, or various business partnerships or other deals.

The network reveals a number of important aspects of the Medici position. The most obvious is that they have more connections than any other family, almost double the number of marriage and business connections of either of their key competitors, the Albizzi and Strozzi.

But beyond the number of connections, the Medici are a key connector of their supporting families, while there is no such unique connector among the opposing families. For instance, none of the Acciaiuoli, Ginori, Pazzi, or Tornabuoni families are directly connected to each other—they all connect via the Medici. This is true well beyond the fifteen families in the figure: out of a fuller data set of ninety-two elite families, more than half of the families married to the Medici were married to at most two other families, while more

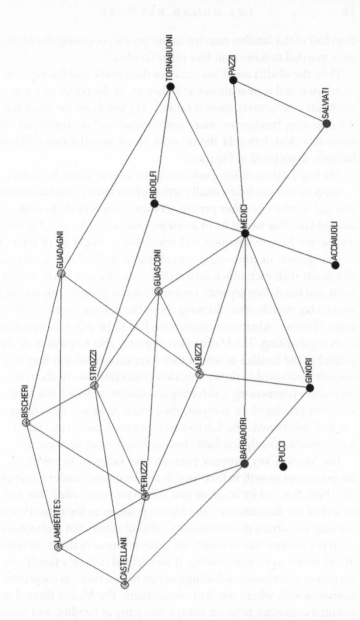

**Figure 2.7: Nodes are some of the key families of fifteenth-century Florence. Solid dark-fill families supported the Medici, hashed-fill families are the opposing families, and the gray-fill (the Salviati and Barbadori) had allegiances to both groups. Links between families indicate a marriage and/or business deal.**

than half of the families married to the families opposing the Medici were married to more than four other families.[36]

Thus, the Medici are at the center of their party, and the key communicator, and appear almost as if they are at the center of a star in their party. In contrast, there is no such key family on the other side: the opposing families are more intermeshed and decentralized. To make this clear, let's split the network above into the two different factions, as pictured in Figure 2.8[37]

The implications of this position can be seen in a simple analogy.

Suppose that you host a small party at which most guests know only you and maybe one other person. Conversation is likely to coalesce around you. You would be in a unique position to know what various people have in common and which topics might be of interest to your guests. In contrast, at a small party at which the guests are all friends with each other, interaction is much more likely to fragment and break into separate conversations or groups. This analogy is not a big stretch when thinking about the politics of early Renaissance Florence, where communication had to be either face-to-face or in handwriting. The Medici were not only able to coordinate the politics of the families to whom they were connected, but they were *uniquely* positioned to do so. The Medici had little worry about their supporters fragmenting and having discussions to which the Medici were not privileged. In contrast, their main rival, the Albizzi, were not well-positioned. The families who opposed the Medici failed to have a leading family and failed to coordinate at key moments.[38]

The Medici's key network position also helps us to understand the enormous growth in their wealth and businesses under Cosimo's direction. It is not by accident that they have been called "the godfathers of the Renaissance." The Medici position in the network was not only important in coordinating political action, but it also made them the obvious intermediary for many business dealings. In order to feel secure in various dealings it was helpful to have a family connection or prior business dealings, or else to deal through some intermediaries with whom one had connections. The Medici formed an essential connector between many other pairs of families, and many of the paths in the network pass through the Medici, many more so than any other family. For instance, just over *half* of the shortest paths between other pairs of families in the marriage network pass through the Medici, while the Strozzi lie on roughly one in ten, and similarly

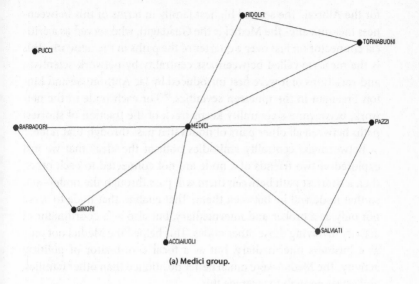

(a) Medici group.

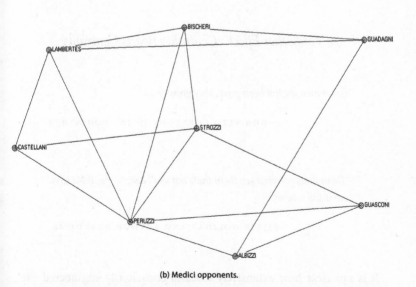

(b) Medici opponents.

**Figure 2.8** Comparison of the networks of families who supported the Medici and those who opposed the Medici. Removing the Medici from their group would completely splinter the network in panel (a), while there is no family that is essential to the network in (b).

for the Albizzi. The second highest family in terms of this betweenness measure after the Medici is the Guadagni, who served as a critical connector on just over a quarter of the paths in the network. This is the measure called betweenness centrality by network scientists, and variations of it were first introduced by Jac Anthonisse and Linton Freeman in the nineteen seventies.[39] For each node in the network, betweenness centrality keeps track of the fraction of shortest paths between all other pairs of nodes that pass through that node.

Betweenness centrality embodies both of the ideas that we just explored: if two friends of a node are not connected to each other, then a shortest path between them will pass through the node—and so that node will lie between them. That enables that node to serve not only as a broker and intermediary, but also as a coordinator of activity involving those other nodes. This helped the Medici not only as a business intermediary, but as a focal coordinator of political activity. The Medici were much better positioned than other families, and astute enough to leverage this.

## A "Godfather Effect": Centrality Begets Centrality

*"Great men are not born great, they grow great."*

—DON VITO CORLEONE IN *THE GODFATHER*

*"Them thats got shall get; them thats not shall lose; So the Bible says, and it still is news."*

—BILLIE HOLIDAY AND ARTHUR HERZOG JR.

It is not clear how extensively Cosimo consciously engineered the Medici's network position, and whether he understood the importance of being a key connector of other families. Nonetheless, the marriages were arranged and not haphazard affairs. Moreover, Cosimo was also known for paying the debts of other families who had been prohibited from participating in Florentine politics because of their indebtedness, and so he actively cultivated connections and the loyalty of others. This was instrumental in the election of the pro-Medici Signoria that enabled Cosimo's return from exile.

However, beyond any deliberate engineering that goes on in the building of networks (more on that in Chapters 5 and 9), there are also feedback effects in how networks form. These feedback effects help explain why some people end up being much more central than others.

Centrality begets centrality. If people gain friends in proportion to how many friends they already have, then friends grow over time like compounding interest. People who are already more central, and wealthier in their number of friends, gain (degree) centrality faster than people who are less central.

In a network setting, such a process has become known as "preferential attachment"—forming new relationships in proportion to how many connections a node already has. Preferential attachment was studied by Albert-László Barabási and Réka Albert and found to generate networks with highly uneven numbers of connections across nodes.[40]

But why would we expect preferential attachment or this sort of compounding in a network setting?

If you want information, it makes sense to connect with someone who is well-positioned in the network. Therefore, someone who has more connections becomes more attractive to connect to.[41] However, there is another aspect to the compounding. Growth in centrality comes not only from how attractive it might be to connect to someone who is central, but also how easy it is to *find* someone who is central. This was something that I explored with Brian Rogers, an economist and former student of mine.[42]

How did you meet your friends? Some you got to know via your friends—you were introduced by a friend or met at a friend's party. The friendship paradox lurks here: the people you are most likely to meet are those who already have the most friends. This leads to a rich-get-richer phenomenon. More central people are easiest to meet, and thus find it easiest to form new friendships.

If you find new friends through existing friends, then the chance of meeting someone is related to how many friends they already have. If someone has twice as many friends as someone else, then they are twice as easy to meet. Knowing any one of their friends gives you a chance to meet them. So, people with more friends will gain new friends at a faster rate. There is evidence that these effects are present in a wide variety of settings, from how researchers find collaborators to how exporters find new business contacts.[43] This is something that is now even being built in to social platforms: you are given suggestions of new people to connect to, or asked "Do you know Person X?," precisely because they are a friend-of-a-friend. The algorithms are suggesting new connections based on the existing network.

As Brian Rogers and I found, the greater the role that the network itself plays in the formation of new relationships, the greater the compounding effect and the resulting inequality in connectedness across nodes. When people find each other via existing relationships, then there can be an enormous multiplier effect. It also increases not only a person's degree, but also other forms of centrality such as eigenvector centrality and diffusion centrality, as they also become more connected to other highly connected nodes. This can be especially important in business settings—where being well-positioned makes one more attractive to attach to *and* easier to find. This makes it easier to gain important contacts and snowballs.

The extent of this effect, and the resulting inequality in network position, varies enormously across settings. Some networks are quite equal, with just a bit of randomness in differences in centrality across people, while other networks are very unequal. The differences in how many close friendships different high school students have look entirely like they arose from flipping coins. In contrast, the variation in how many links point to different Web pages is much more unequal, and looks as if they were found predominantly via their existing links—the easier they are to find by following existing links, the more links they gain. There is much greater inequality in the degrees across Web pages than across close high school friendships.[44]

## A Taxonomy of Networked Influence and Power

We have seen that network structure provides essential insights into influence beyond measures of wealth and political power. Moreover, network structure is often important beyond a simple count of how many connections each member has. Our taxonomy has four basic ways in which people wield influence, with associated measures of how "central" a person is in their network:[45]

- Popularity—"Degree Centrality": Does someone have many friends, acquaintances, and followers? Being able to get a message out to millions of followers on a social medium gives a person the potential to influence what many people think or know. Popular individuals are disproportionately observed, and hence can bias people's views of trends and norms.
- Connections ("It's Who You Know")—"Eigenvector Centrality": Is a person connected to other "well-connected" people? Having many friends can be useful, but it can be equally or even more important to have a few well-positioned friends.
- Reach—"Diffusion Centrality": How well positioned is a person to spread information and to be one of the first to hear it? Can a given individual reach many others within a short number of hops in the network?[46]
- Brokerage and Bridging—"Betweenness Centrality": Is someone a powerful broker, essential intermediary, or in a unique position to coordinate others? Must others need to interact through this individual in order to reach each other? Does an individual serve as a key bridge from one group to another, connecting otherwise disconnected groups?

There is a sense in which popularity is a local measure—one just needs to count a person's friends or acquaintances, while the other three concepts are more holistic, embodying information about larger parts of the network. Which centrality concept is appropriate, and how people wield power, depends on the context; and these concepts play important roles in contagions, inequality, and polarization, as we shall soon see.

## 3 · DIFFUSION AND CONTAGION

*"How many valiant men, how many fair ladies, how many sprightly youths, . . . , breakfasted with their kinsfolk, comrades and friends, and the same night supped with their ancestors in the other world!"*

—GIOVANNI BOCCACCIO, *THE DECAMERON*, 1353

The bubonic plague, or Black Death, spread across Europe, slowly but steadily, from 1347 to around 1352.

The culprit, *Yersinia pestis*, is a pathogen carried by fleas who ingest it when feeding on an infected host. It blocks the fleas' intestines causing them to become starved for nutrients, which leads them to feed voraciously and infect their subsequent hosts. Fleas are adept at living on rats, other animals, and humans; with some resistant hosts serving only as carriers and others quickly dying once bitten and infected. It is a horrifying disease: beginning like a flu with weakness and fever, but turning to extensive hemorrhaging. The dying tissues turn black, giving the plague its nickname of Black Death.

The sanitation of the era, a lack of understanding of contagion, and close proximity of humans and many animals meant that the disease was amazingly virulent in the growing cities of the Middle Ages.[1] It cut the populations of Paris and Florence roughly in half within a couple of years, with even larger death tolls in cities like Hamburg and London. It is believed to have made its way along the Silk Road from China to Constantinople, and later from Genoese trading ships to Sicily by 1347, where it quickly wiped out roughly half of the island's population. It continued to spread, hitting parts of Italy, and then Marseille, before spreading through France and Spain, and eventually getting to the northern countries a few years later. Overall, it is estimated to have killed more than 40 percent of Europe's population, as well as 25 million people in China and India before even reaching Europe.

What is remarkable from a modern perspective is how *slowly* and methodically it spread. Although the plague did make occasional long-range jumps, as in its travel along trading routes such as the Silk Road and via ships, its progression throughout Europe averaged only about two kilometers per day, slow even by the standards of foot travel at the time.[2] Even though the bubonic plague rarely transmits directly from person to person, the disease traveled alongside humans—via the fleas who fed on rats on ships, on farm animals, people, and in clothing—and so it made its way through the networks of humans and the various animals that accompanied them.

The slow movement of the plague tells us how limited the mobility and range of contacts of most humans was in the Middle Ages. Modern pandemics are quite different: they spread remarkably quickly, with diseases jumping continents typically within a matter of days or weeks. A measles outbreak among unvaccinated adults and children sparked via interactions at an American theme park in southern California in 2014 appeared in schools hundreds of miles away days later. Ebola was carried by health workers from Sierra Leone in 2015 to cities in Europe and North America within a week of their exposure.

In this chapter we will see how contagion and diffusion depend on the structure of our networks. Beyond immediate insights into the spread of diseases, this understanding will also serve as a starting point for comprehending the more complex spread of ideas, financial contagions, and inequality in employment and wages—topics of some of the following chapters.

## Contagion and Network Components

*Lycus: Is it contagious?*
*Pseudolus: Have you ever seen a plague that wasn't?*

—BURT SHEVELOVE AND LARRY GELBART,
*A FUNNY THING HAPPENED ON THE WAY TO THE FORUM*

Although there are big differences between many of our networks and those of the Middle Ages, we can still learn much about the slow but relentless spread of the plague by looking at a particular type of modern network.

Figure 3.1 pictures a network of romantic and/or sexual relationships among teenagers in a U.S. high school. The students listed their liaisons over eighteen months.[3]

Even though a typical individual in the network in Figure 3.1 has only one or two interactions, the network still exhibits a "giant component": the large connected piece in the upper left of the figure in which 288 of the students are connected to each other via sequences of relationships.

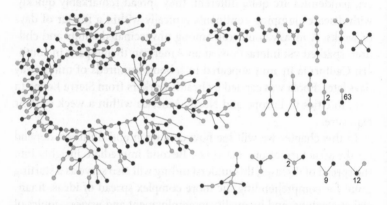

Figure 3.1: A network from a high school in the Midwestern United States from the Add Health data set. Nodes are students, colored by gender. A link denotes a romantic or sexual relationship during an eighteen-month period. The numbers by some of the components indicate how many times that component appears (e.g., there are sixty-three couples who only had a relationship with each other). Isolated students are not pictured. Just over half of the students are in one giant component on the left. The data in this figure were first analyzed and discussed by Peter Bearman, James Moody, and Katherine Stovel (2004).

"Components" are the pieces of a network in which each node can reach each other via a path of connections.[4] Just over half of the students in the figure are in the giant component, and the rest sit in many small components.[5] More than a quarter of the students reported no relationships (we all remember how lonely high school can be) and are not pictured.

This figure highlights how a sexually transmitted disease can infect a large fraction of the population, even though each individual has only a few interactions on average. Each link represents a potential for spreading a disease from one individual to another. If someone in the giant component were to become infected (for example, via an interaction with someone outside of the school), then the disease could spread widely within the giant component and thus within the school.[6]

As an example, HPV (the human papillomavirus) is sexually transmitted and can lead to several cancers, including cervical cancer. A danger with HPV is that it is often asymptomatic so that an infected person has no reason to believe that they are infected and so may continue to spread it to others. More than 40 percent of the adult population of the United States is estimated to have HPV, many unaware.[7] Most of those infected are not promiscuous; they just happen to be part of the giant component.

From Figure 3.1, it is easy to see how a disease could spread slowly, given the relatively low number of contacts per individual, but could still eventually lead to a high level of infection as it spreads throughout the giant component, just as it did with the bubonic plague.

We also see from this figure that a disease's spread is not dependent on the presence of highly promiscuous individuals or sex workers. High-degree individuals can amplify and accelerate the spread of diseases, but they are not necessary for a network to have a giant component. Simply having more than one interaction per individual is enough.

This network sits right at the juncture of connectivity at which widespread contagion becomes possible.

## Phase Transitions and Basic Reproduction Numbers

The term "phase transition" is often used in thermodynamics to refer to changes in matter.[8] For instance, as water changes to ice or to steam it is said to undergo a phase transition.

Networks also undergo phase transitions, from being collections of isolated nodes and small components, to a network that has a giant component containing a nontrivial fraction of nodes, and then eventually to one in which all nodes can reach one another via paths in the network. Increasing the fraction of links present in a network is an analog of increasing the temperature and changing ice to water to steam.

The remarkable thing about phase transitions is how abrupt they can be. Just below the freezing threshold you are standing on ice and yet only a degree higher you are plunging into the water. Similarly, tiny changes in the frequency of links in a network have dramatic effects on its component structure. This is illustrated in Figure 3.2. As we move from one half of a friend per person on average (as in panel [a]) to one and a half friends per person (as in panel [b]), we transition from a network that is disconnected to one in which a majority of people can reach each other. Small further increases (panels

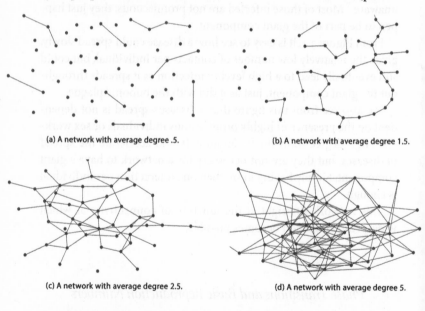

(a) A network with average degree .5.                    (b) A network with average degree 1.5.

(c) A network with average degree 2.5.                    (d) A network with average degree 5.

Figure 3.2: A comparison of networks with varying average degrees. With less than one connection per node, as in panel (a), the network is fragmented. Once there is more than one connection per node on average, as in panel (b), a giant component coalesces—the nontrivially sized group of nodes at the bottom of panel (b) in which all can reach each other via paths in the network. Slight additional increases in the connections per node lead the giant component to involve almost all nodes, as in panel (c), and eventually lead the network to become path-connected so that every two nodes have a path between them, as in panel (d).

[c] and [d]) lead the network to become "path-connected," or "con. nected" for short: each person can reach every other via paths in the network (panel [c] is just on the verge, with two nodes left out).

Phase transitions in networks are fundamental to fighting disease. A critical number associated with a disease and a network through which it might spread is known as the disease's "basic reproduction number." This tracks how many other people are newly infected by a typical infected individual. If its basic reproduction number is above one, then a disease spreads, while if it is below one, then the disease dies out.

The threshold of having a basic reproduction of one corresponds to the phase transition at which networks have a giant component, as in Figure 3.2. The idea behind this is simple but vital: with more than one new infection per infected individual, the contagion continues to expand, reaching more people with each new infection, and so can perpetuate itself. Below that level, the process dies off. In terms of the network, if each person has more than one friend, then a component tends to grow outward and expand to be a giant component, while with fewer than one friend on average, the network is a bunch of small disconnected components and isolated nodes. The analogy to reproduction is clear: if a society has more than one child per adult (who then survives to reproduce), then that society will grow; while having fewer than one child per adult leads a society to shrink.

It is easy to find examples of the extinction or near-extinction of a population as its reproduction dropped below one surviving offspring per adult, and where that reproduction number depends on circumstances. The American bison is thought to have numbered more than fifty million in the eighteenth century and was down to five hundred by the end of the nineteenth century. Their reproduction number plummeted after the U.S. Civil War, as new train lines brought more hunters to the herds and made it easier for them to transport their hides. Better guns also allowed hunters to kill animals at great distances without frightening a herd. For instance, the "Big Fifty" that was developed in the 1870s by Sharps Rifle had a reliable range of over a quarter mile (more than four hundred meters). The Plains Indians called it the gun that "shoots today and kills tomorrow."[9] The growing numbers of hunters, each killing more bison with improved rifles, and transporting them more quickly, led bison to be killed at a rate much faster than they could reproduce. The bisons' reproduction number abruptly dropped, and the existing population was all but eradicated in a few decades.

The basic reproduction number of a disease depends on how easily it spreads from one individual to another, as well as on with how many people each individual has contact. Since not every contact transmits a disease, the basic reproduction number is generally lower than the average degree of people in the network. Thus, reproduction numbers differ across diseases and locations.

Ebola's basic reproduction number (in the absence of intervention) has been estimated to be just over 1.5 in Guinea and Liberia, but closer to 2.5 in Sierra Leone.[10] This difference stems from differences in population densities, which affect the average number of people that a person has contact with per day, with Sierra Leone's being more than 60 percent higher than that in Guinea and Liberia.

The measles' reproduction number, in contrast, is much higher than Ebola's since instead of spreading via blood and saliva, it spreads via airborne particles and has a reproduction number from 12 to 18 depending on local population densities and interaction frequencies. Measles are very dangerous in unvaccinated populations. Diseases such as diphtheria, mumps, polio, and rubella, are intermediate, in the 4 to 7 range.[11]

The differences in these numbers correspond to different networks. HIV (human immunodeficiency virus) spreads via intimate contact, whereas one can catch the flu from a handshake or sitting near a coughing person on a plane or bus. That leads to many more interactions in the network of a flu, and fewer connections in the network of HIV. This does not mean that HIV does not spread: its reproduction number in some parts of the world and among some subsets of the population is well above one, and so it is still endemic among many communities around the globe.[12]

Reproduction numbers lie at the heart of vaccination policies. A vaccine does not need to be fully effective or to reach every individual in order to avoid widespread contagion, it just needs to bring the reproduction number below one. Vaccinating individuals not only keeps those individuals safe, but it also eliminates their connections from the network. Thus, it lowers the reproduction number of the society and helps protect the remaining population. If we start with a reproduction number of two, so that each infected person would infect on average two others, then vaccinating just over half of the individuals would drop the reproduction number to below one and limit the spread of the disease.

Unfortunately, the incentives that people have to vaccinate themselves are part of the reason that diseases are so difficult to eradicate. Those incentives are suboptimal because of what are known as "externalities."

## Externalities and Vaccination

*"Thousands of candles can be lighted from a single candle, and the life of the candle will not be shortened. Happiness never decreases by being shared."*

—THE BUDDHA

*"It may easily happen that the benefits of a well-placed light-house must be largely enjoyed by ships on which no toll could be conveniently levied."*

—HENRY SIDGWICK,
*THE PRINCIPLES OF POLITICAL ECONOMY,* 1883

Henry Sidgwick was born in 1838 in Yorkshire, the year after Queen Victoria began her reign in England, and died in 1900, one year before Victoria. He was known for many things during his lifetime besides being one of the first to really pinpoint externalities.[13] He played a role in debunking psychics, including one of the more famous of the day—the medium Eusapia Palladino. Sidgwick was also the founder of Newnham College, the second college for women to be part of the University of Cambridge. He wrote essays in moral theory, which had many of its foundations laid during the Victorian era.

For us, however, Henry Sidgwick's legacy lies in his quote above, which illuminates the concept of externalities: *one person's behavior affects the well-being of others.*[14] In Sidgwick's quote it is a ship benefiting from the presence of a lighthouse that someone else built and maintains.

We have all experienced externalities in the small and large: having a neighbor learn to play drums, having someone kick our seat on a long plane ride, or sitting in a traffic jam. And, as climate change illustrates, externalities can even extend to people yet unborn—as future generations will experience a climate that is in part determined by our emissions.

Now that you are familiar with the concept, you will notice externalities everywhere. They make human interaction interesting and externalities prevent free markets from being a panacea. Externalities lie at the heart of moral and ethical quandaries, as well as many of the most pressing social and economic problems, ranging from freedom of speech to gun control and climate change. As externalities are fundamental to networks, they will keep reappearing in this book.[15]

When a worker in a coffee shop in an airport gets a vaccination against the flu, it not only helps him or her stay healthy, but also helps the many travelers who might otherwise have been infected if that worker caught the flu. The externality is that the worker's decision of whether to get a vaccine ends up affecting whether other people get sick. The worker might not fully take all those other people's potential suffering into account when making his or her vaccination decision. Stanford University, as do many organizations, understands this and tries to help people make the right decisions and so provides free flu vaccines for its staff and students. The vaccination of even a part of a community conveys benefits to the whole community. Governments pay special attention to the vaccinations of schoolchildren, teachers, health workers, and the elderly—categories of people particularly susceptible not only to catching but also to transmitting a disease.

It's not accidental that governments are heavily involved with vaccination. When there are externalities, free markets fail to align individual incentives with society's overall well-being. A parent weighing the costs and benefits of a vaccine for their child is not always thinking of the broader consequences of that vaccination to other people. These are markets in which subsidizing or regulating behavior can make everyone better off. The reason for requiring that a child be vaccinated before enrolling in school is not just to protect that child, but because each child's vaccination affects others via potential contagions. Small pockets of unvaccinated individuals can allow a disease to gain a toehold and spread more widely.

The biggest challenge in eradicating a disease is that externalities operate on a global scale. China was declared free of polio in the year 2000, but then in 2011 had an outbreak that appears to have made its way in from a neighboring country. Great strides have been made in the fight against polio, given that it was present in more than one hundred countries as recently as 1988. However, even having one country in which a disease is endemic is enough to keep it alive and allow it to resurge and spread again to other countries. Keeping a population vigilant against diseases that have seemingly disappeared is costly and challenging. It can be incredibly frustrating to have to keep vaccinating children around the planet year after year simply because a couple of countries are delinquent and keep incubating a disease.

Vaccination policies also have negative feedbacks: the more successful a vaccination effort is, the lower the threat of the disease and the lower the incentives for the population to remain vigilant. When a disease is running rampant, people pay attention and vaccinate themselves—not because of a concern about the externalities and others' health, but because they become scared for themselves. Deadly outbreaks of smallpox led to some of the first inoculations: centuries before vaccinations were formally developed, people in China were taking bits of dried pox from victims and either inhaling them or scratching them into their skin to gain immunity. However, once a disease subsides, people lose their fear and vaccination rates fall—leading the reproduction number to grow and allowing the disease to resurge.

This feedback effect can lead to especially strong cycles since many people fear vaccinations (more on that in Chapter 7) and so avoid vaccinations whenever a disease becomes less visible. Given the abrupt phase transition in a disease's reproduction number with small changes in vaccination rates, and the global scale of contagion networks, it becomes hard to eradicate any disease, and most tend to cycle over time. Smallpox is the only human disease that has been officially eradicated according to the World Health Organization (WHO). The last recorded wild case was in 1977 in Somalia, and in 1980 WHO said that the disease was eradicated. Fully eradicating smallpox was no small feat, as it involved decades of quick response to any observed new outbreak and then isolating patients and quickly vaccinating people in the area.

## Well-Connected but Sparse

The good and bad news about human networks is that many of them are *well-connected:* having most people in a giant component. Although connected networks pose a challenge in controlling disease, they are vital in the spread of useful information, for instance, about a despotic government, an exciting new book or movie, or a valuable new technology.

Interestingly, human networks are well-connected even though they tend to be sparse at the same time. This sounds like a contradiction, but let me explain.

Consider Facebook. According to a recent Pew Research Center survey,[16] adult users on Facebook in the U.S. have an average of 338 friends and more than half of all adult users have over 200 friends. The numbers of friends among teen users are higher. This puts us well beyond the threshold of one friend per person at which networks begin to become connected. In that sense, human networks such as Facebook are extremely well-connected. Indeed, *99.9 percent* of Facebook's more than 700 million active users are in a single giant component.[17] So, except for a few isolated individuals and small groups, almost all Facebook users in the world can have information reach them from almost any other user via paths of friendships on the platform.

If almost everyone in the network is in one giant component, how is Facebook's network sparse? "Sparsity" refers to the fact that you could hypothetically have up to 720 million friends on Facebook, but you don't. We all know people who have thousands of friends on Facebook (don't forget the friendship paradox!), but nobody comes close to having even a percentage of all the possible friendships they could have. Having only hundreds of friends on average out of a potential hundreds of millions means that fewer than one in a million possible friendships on Facebook are actually present. The Facebook network has a minuscule fraction of its possible links present and hence is extremely sparse. Yet that tiny percentage of actual links is enough to bring almost all of the users into one giant component.

Beyond having almost all users in a giant component, despite the sparsity of Facebook's network, the paths between users are extremely short. Perhaps astoundingly, the average distance between any two active users is only 4.7 links.[18] This is known as the "small-world" phenomenon. It has a popular history via a Hungarian short story written in 1929 by Frigyes Karinthy, and later by John Guare's play *Six Degrees of Separation*. The small-world phenomenon is a robust feature of many random networks as discovered by a series of mathematicians in the 1950s.[19] It also played a starring role in an important book, *Small Worlds,* by Duncan Watts (1999).

The small-world phenomenon was beautifully illustrated in experiments conducted in the mid-1960s by psychologist Stanley Milgram. Milgram's starting subjects were people who lived in Witchita, Kansas, and Omaha, Nebraska, who responded to a letter Milgram sent out to residents asking them to participate in a study. Those people were asked to get a folder to some target individuals in Massachusetts. The targets were people Milgram had selected to help him with the experiment. One target was a stockbroker and the other was the wife of a divinity school student. The subjects in the experiments were told the targets' names and the towns in which they lived and a bit about them. The instructions to the subjects in the experiments were: "If you do not know the target person on a personal basis, do not try to contact him directly. Instead mail this folder . . . to a personal acquaintance who is more likely than you to know the target person . . . it must be someone you know on a first-name basis." Each person who received the folder read the instructions and added some of their personal information to the folder and then sent it along.

One folder started with a wheat farmer in Kansas. The farmer sent it to a minister in his hometown. The minister then sent it to a minister he knew in Cambridge, Massachusetts, who happened to know the target stockbroker directly. In this case, the folder went from its starting person in Kansas to the target across the United States in just three steps.

After collecting the folders from the targets, Milgram could see how many folders made it to their destination and how many steps it took each folder to reach the target. Out of the 160 folders that started in Nebraska, 44 made it to the final target—27.5 percent. The median number of steps was five and the range was from two to ten, with an average just above five.[20]

Given that all the people who had folders sent to them along the way were not volunteers for the experiment, but instead simply received the folder from an acquaintance, one might expect them to have a fairly low chance of forwarding the folders. Thus, the percentage reaching the final targets is impressive. However, the fact that participation was voluntary also means that the low number of steps of the folders that reached their target partly reflects a bias in the experiment. If a folder would have to take a longer path, involving ten people sending it along rather than five, then twice as many people would have to participate in order for it to make it. This makes it much more likely that paths with small numbers of intermediaries are successful and appear in the data, while ones that would have required more intermediaries are more likely to fail. Later experiments that correct for that bias find averages on the order of ten hops, double Milgram's results, but still relatively small.[21]

The results of the experiment are remarkable not only because of how few hops they took, but also because many letters made it at all despite the fact that people didn't have any map of the network to guide them in forwarding the folders. It would only be by the wildest chance that you would happen to know the shortest paths in the network between you and some stockbroker in Massachusetts, or, in later experiments, to a student in Beijing, or a plumber in London, and so forth. Thus, the fact that many of these folders were passed along fairly short paths suggests not only that short paths exist, but that many short paths tend to exist between any pair of people, and that people know enough to figure out how to pass something along fairly efficiently. How people are able to navigate a network is something that we will come back to in Chapter 5.

How is it that a network can be so sparse, having less than one in a million of its links present, and still require only a handful of links to get from any person to any other of the hundreds of millions of users?[22] Let us take the Facebook network as an example, and work with a typical user, say Diana, in terms of number of friends. A typical Diana would have a few hundred friends, and let's take the average at roughly 200 friends with whom she interacts at least occasionally.[23] Now let us count Diana's second-degree friends—people who it takes two links to reach from her. Let us suppose that each friend again has 200 friends that are not already Diana's friends.[24] Thus, by moving out paths of length 2 we have reached 200 × 200 = 40,000 users. Continuing, we reach 8 million people in three steps and 1.6 billion by the time we have gone out 4 steps. We have more than covered the full population of Facebook. Moreover, most of the users are reached at the later steps—most users are either 4 or 5 connections away from each other. This gives us the idea of why human networks have such small distances between people.

## Our Ever-Shrinking World

*"The pilgrims didn't know it, but they were moving into a cemetery."*

—CHARLES C. MANN[25]

Let us compare this modern world network with one from medieval life. Suppose that instead of having 200 friends, we do the same calculation with 5 friends. After four steps we would have reached roughly 5 × 5 × 5 × 5 = 625 people instead of 1.6 billion. To reach the world population of the day would require more than a dozen steps instead of four or five.[26]

Nevertheless, the medieval world was still largely connected—as even a few friends put us above the reproduction number of one. And even the medieval world had a small-world aspect to it. Typical distances of a dozen or more links needed to get from one person to another are larger than the modern four or five, but still small compared to the hundreds of millions of people that were alive then. The greater distances of medieval times did lead to slower and more sporadic travel of germs and ideas than we see today. Yet the world was connected enough for long-range transmission and contagion, as we see from the relentless spread of a long list of diseases that made human survival a constant battle.[27]

Once global travel started to involve hundreds of thousands of people, the world began to see very fast and deadly pandemics. An eye-opening example is the 1918–1919 flu season. The flu of that season was a particularly nasty strain: it was unusually deadly among young and otherwise healthy populations, as it led to an overreactive immunity response that resulted in deaths of more than 10 percent of those infected. It became known as the Spanish flu, which was a disservice to the Spanish. The Spanish were being accurate in reporting infection and mortality rates, while information was being suppressed in other countries to maintain morale after the devastating world war of 1914–1918.[28] The news made it appear as if it was an epidemic coming from Spain, even though it was already widespread. The key to the spread of the flu that year comes from the end of the war, which led to mass troop movements around the world. Many soldiers were living in tight quarters and traveling great distances. This was coupled with a disease that has two features that enable it to spread quickly and extensively through human populations. One is that the flu can be communicated via small droplets that become airborne when someone sneezes or coughs and can travel from one person to another at a distance of over a meter, and can also be left on surfaces to be touched by someone else. The second is that people can be contagious for periods of over a week, sometimes beginning before symptoms emerge and ending after symptoms have subsided. The combination of a nasty flu, no vaccinations, and large masses of people moving around the world led to one of the largest flu pandemics in history and with deadly consequences. The flu infected on the order of a half a billion people (about a third of the world's population, and much more in urban Europe), and claimed somewhere between 50 to 100 million lives around the globe.

This example also points out that human networks are not constant in their connectivity. The mass troop movements of that year were unusual. They led to a smaller world than in previous years. Beyond occasional dramatic changes in human travel, there is a strong seasonality in how much people interact. For instance, the seasonality of school openings drives spikes in various diseases. This was first documented in 1929 by Herbert Soper, a statistician who studied the fluctuations of many diseases over time. He noted that measles outbreaks in Glasgow had patterns that could be explained by school sessions. When school is in session, many children who lack immunity to various diseases are in close proximity with each other, and so the connectivity of the network on a local level is quite high. In contrast, during times in which schools are closed, that local connectivity drops dramatically. However, longer trips and travel during breaks lead to increased long-distance connectivity.[29] Thus, networks of interactions change in more than one way depending on the season. Modern epidemiological models that are used to predict the spread of diseases, especially those such as the flu, take into account school seasons, travel patterns, interactions with health workers, and many other factors that affect the connectivity of the networks of transmission.

The deadliness of transported contagions was never more dramatic than the introduction of smallpox, measles, typhus, and influenzas to the Americas. It is estimated that those diseases ultimately have killed more than *90 percent* of the native population.[30] The native American populations were varied in their densities and the degree to which they interacted with each other and so it took time for the devastation to spread.

In Mexico, the arrival of smallpox on a Spanish ship from Cuba in 1520, via an infected slave, managed to devastate the majority of Aztecs in a matter of years. Within a decade it had made its way to eradicate most of the Incas in South America. Epidemics would also sweep through North America, killing large populations in the fertile zones in the East and Midwest, which were relatively densely populated. Some of the more remote and less dense populations in North America would last another century before their exposure, but none escaped. Native Americans in parts of the New England coastal area where the Pilgrims landed were devastated just a couple of years before the Pilgrims arrived. With much of the competition for land and resources decimated, the Pilgrims had a much better chance at survival than if they had encountered the denser native American population that existed just a few years prior.

Some of the last to be killed were the native Hawaiian populations, who lasted until the nineteenth century before they were eventually visited and then repeatedly battered by Eurasian diseases. The voyage of King Kamehameha II and Queen Kamamalu to London to negotiate a treaty led to their demise, as well as most of their party. They contracted the measles when they visited the Royal Military Asylum, which was full of soldiers' children.[31] Measles would eventually make their way to Hilo, Hawaii, in 1848 via a U.S. Navy frigate, the *Independence*, coming from Mexico.[32] Whooping cough and flus would make their way to Hawaii that same winter, starting a series of epidemics together with the measles that would eventually conquer roughly a quarter of the native population. The year would be called the "year of death" in the census. Before the society could get back on its feet, smallpox was delivered in 1853 via another ship, the *Charles Mallory,* which sailed into Honolulu from San Francisco. The ship was thought to be effectively quarantined and eventually sailed, but it had left the disease behind and within months thousands had died. When Captain Cook first arrived in Hawaii in 1778, the local population was estimated to be over 300,000; but fewer than 40,000 native Hawaiians made it to the 1900 census.

Modern medicine has greatly improved the understanding of contagion and the importance of sanitation and vaccination, and reduced the day-to-day threat of many diseases. Although we are far from eliminating pandemics, it is impressive that humans still survive despite the world becoming increasingly interconnected. The number of other people with which a typical individual in the industrialized world interacts is orders of magnitude larger than it was a few centuries ago, especially as we regularly rely on many others for our food and sanitation. Moreover, modern travel means that many interactions occur across large distances—with hundreds of thousands of people traveling internationally on any given day. Thus, potential contagion networks for many diseases have three big differences from the high school relationship network in Figure 3.1: they are denser, they include almost all nodes in the giant component, and they have shorter average distances between nodes. This means that the potential for many contagions to spread both rapidly and widely is much greater today than it was centuries ago, when such pandemics repeatedly wiped out millions. We can hope that science and the development of new vaccines continues to outpace the appearance of new diseases and the increase in the human network's connectivity.

## *Centrality and Contagion: The Downside of Popularity*

The friendship paradox that we discussed in measuring centrality and influence also has implications for contagion and diffusion.

Being relatively overrepresented among people's friends not only gives someone high influence, but also high exposure. So, if you are at times jealous of your friends' high popularity, here's your silver lining. The most popular can be among the first to hear new news but also the first exposed to new infections.

A notorious example of this was a Canadian flight attendant, Gaëtan Dugas. A Centers for Disease Control study found that by 1983, out of the 248 people who were known to have HIV at that time, 40 had sexual contact with Dugas. Much was made about Dugas being "patient zero," and he was widely blamed for the epidemic that ensued.[33] With more data and hindsight, it is clear that AIDS had actually made its way into the U.S. by the 1960s, most likely via Haiti (and originally from Africa where it has even earlier roots), and that it would have become well-entrenched in the world without a promiscuous flight attendant. Nonetheless, Dugas helped stoke the most noticed early outbreak.

Similarly, it has been estimated that as few as 3 percent of people infected with Ebola may have spread more than half of the cases in an outbreak in Sierra Leone. Again, the outbreak would have occurred without highly connected people, but they are more exposed and can accelerate the spreading.[34]

To see why high-degree people are not vital for epidemics, it is enough just to look at a network. Again, revisiting our Figure 3.1, we saw a large giant component and yet the network has very few people with high degrees: only one person with degree 7, one with 6, and a handful with degree 5, and most nodes with degree 1 or 2.

This is important to stress, as it is a common misconception about networks. Hubs and connector nodes are not always *necessary* for a network to be connected and host contagions or diffusion. They may be more prone to be involved, and may provide early sparks, but many contagions would occur even without the most connected nodes. If we eliminated a few of the highest-degree nodes in the romance network, we would separate a few small bits from the giant component, but it would still be largely intact. The driving force behind a giant component that harbors widespread infection is the overall average degree in a network. The tendency for most individuals to have a degree above one in many human networks is what makes contagions and the diffusion of information so ubiquitous.

Nonetheless, high-degree individuals are more susceptible to infections, can accelerate transmission, and in networks that are right at the phase transition, can make a difference. More important, if one wants to target nodes that might have the biggest impact, the most central nodes are the place to start. As such, the idea that higher-degree individuals are at higher risk of infection has helped guide new analyses of the spread of diseases in the wild.

As an example, Stephanie Godfrey and her colleagues in Australia and New Zealand studied how prevalent ticks and mites are in populations of tuatara—a lizardlike reptile that lives in New Zealand.[35] The tuatara were named by the Maori for the ridges on their back. Tuatara are actually a fascinating species—not truly lizards, but instead the last remnant of the Rhynchocephalia, of which the other species became extinct more than sixty million years ago at the end of the Cretaceous period, along with many dinosaurs. The tuatara have a third "eye" on the top of their heads that is not used for vision, but is hypothesized to absorb ultraviolet rays and regulate the tuatara's metabolism. Tuatara live a fairly lonely life, spending most of their time alone within their own territory eating bugs as well as an occasional bird egg or frog, and basking in the sun. Not a bad life, at least on some dimensions.

Figure 3.3: A tuatara[36]

Despite their solitary nature, their territories overlap and thus tuatara come into occasional contact with each other—and, of course, their reproduction requires it. The tuatara are hosts to a form of tick that carries a blood parasite, harmful to the tuatara, and the tuatara also are often infected with a type of mite. The interesting aspect from a network perspective is that the ticks don't live off of the hosts for long. Thus, moving from one tuatara to another requires the tuatara to come in close contact and so the network of interactions is important for the spread of ticks. In contrast, the mites can live off of the tuatara, and so are less dependent on tuatara interactions to spread.

Godfrey and her colleagues followed many of tuatara on Stephens Island and charted their movements and territories. Their territories had very different patterns, so that some tuatara only overlapped with one other tuatara, while some were much more central and overlapped territories with ten or more others. This is their degree centrality: how many other individuals they had a chance to interact with. Then by counting ticks on the tuatara (how did you spend your summer?), Godfrey and her colleagues found that there was a substantial and significant correlation between the degree centrality of a tuatara and how many ticks, and associated blood parasites, it had. Since ticks depended on the network to go from one tuatara to another, having a high degree put a tuatara at greater risk. Interestingly, they did not find the same relationship for the mites that can survive off of the tuatara. Here, the network was not essential to transmission, and the degree did not matter in the infection rate.[37]

Such analyses have been conducted for a variety of species,[38] including humans. Nicholas Christakis and James Fowler[39] examined which students at Harvard University came down with the flu the earliest. They monitored two groups: one group of a few hundred students who were picked at random from the population, and another group of a few hundred students who were named by others as a friend. As we know from the friendship paradox, the students named as friends should have higher degree than those picked randomly from the population. Indeed, as Christakis and Fowler found, those named as a friend by others had the flu on average two weeks earlier than the random group of students. Popularity has its downside.[40]

## Network Dynamics and Conductance

In 2009 an unusually deadly and dangerous flu strain, a variety of H1N1 virus, spread around the world—a close relative of the virus responsible for the Spanish flu that devastated the world's population in 1918.

Along with all of the other people flying into Beijing that summer, I walked by a device that took my temperature. China was not the only country to screen travelers. Dozens of countries screened travelers and asked them to fill out forms reporting any symptoms.

People who were thought to be infected were denied entry or quarantined. The network was changing in response to the disease.

In some cases, the travel restrictions and alerts turned out to be extremely costly. As Mexico had some of the first H1N1 flu cases in 2009, many of the travel alerts that were issued that spring mentioned Mexico. That led travel to and from Mexico to drop by around 40 percent in the late spring of 2009. For a country in which tourism is a major industry, such an abrupt and huge drop in travel was deeply felt.

In hindsight, by carefully analyzing networks of travel, as well as the timing and location of cases of flu around the world, we can see that the change in travel did little to stem the spread of the flu. Travel changes delayed the spread by a few days.[41] Even the countries that did the strongest travel screening look to have only delayed the flu from spreading widely within their border by seven to twelve days, and did not avoid the inevitable contagion.[42]

World travel is so extensive these days that even cutting a large portion of it, and catching as many infected individuals as possible, makes a small difference in the spread of a flu. We can think of such strategies as cutting some, but not all, of the many connections that move long distances in the world network. It does not come close to really undercutting the reproduction number of such a flu. Of course, this does not mean that an individual could not remain healthy by avoiding travel during a flu pandemic. If you want to spend the flu season in a cabin in the remote mountains, you can all but eliminate your personal chances of catching the flu. But cutting the travel of large populations is economically infeasible.

Attempts at quarantining have on occasion even been disastrous, especially before contagion was well-understood. Reactions to early polio epidemics illustrate this point. Polio had been around from at least the days of ancient Egypt, and its infected included many famous people, from Emperor Claudius to Sir Walter Scott, but it often popped up fairly randomly. It began to appear in larger epidemics around 1910 in Europe, and the polio epidemic that hit New York in the summer of 1916 was large and dramatic. Polio was ill-understood at the time: children would go to bed one night and wake up in the morning unable to walk.

The epidemics were terrifying and not surprisingly led to panic.

Polio is transmitted from human feces to other humans orally: so, having open sewers near children is a deadly mix. But the variety of hypotheses surrounding polio led to the killing of eighty thousand cats and dogs, and in addition people blamed mosquitoes, mercury, bedbugs, and many other things. The majority of the first cases in New York happened to be Italian and so some Italian neighborhoods were quarantined. The quarantining led sanitation to deteriorate and more children to become exposed; and children who developed fevers for other reasons were shut in with others who had polio, with deadly consequences.[43]

This does not mean that changing the contact patterns in a network is never an effective strategy. With a disease like Ebola, with a much lower basic reproduction number, identifying outbreaks at an early point and restricting travel in and out appears to have been effective. This is also aided by the fact that the outbreaks have often been in places with lower rates of travel. Restricting travel around a village in Sierra Leone is different from trying to cut travel in and out of Beijing, London, New York, or Mexico City. A variety of studies[44] suggest that the only ways to effectively manage large flu pandemics are by vaccinating, quarantining infected individuals (making sure they stay at home or in a clinic until no longer contagious), and in some cases using antivirals that shorten infection and lower the chance that it is transmitted. These methods can all significantly lower the reproduction numbers of a flu and have a substantial impact.

The point here is that networks change and react to what passes along their connections. With the spread of dangerous contagions, such as diseases or financial distress, people react with fear, cutting ties, isolating nodes, and turtling up. In the other direction, the arrival of some important news can lead people to actively contact each other and increase a network's density—accelerating the spread of good news and salacious rumors. Fully understanding the contagion properties of a network depends on understanding that networks are dynamic entities and they often react to a contagion. We will return to some of these ideas in Chapters 7 and 8, where we will discuss things like technology adoption, decisions to invest in education, and social learning. Those are processes in which the way people act is dependent on what others are doing and the state of the network.

## Collecting Thoughts

In the modern world, many of our networks are connected and, for better and worse, you sit in the giant component, along with most of the rest of us. We are constantly exposed to flus and other diseases, but also privy to the spread of the latest news and rumors. Some news is almost impossible to avoid hearing.

As part of a fun diversion, a group of people challenge themselves to have low numbers of interactions and to be the last to hear a piece of news. The challenge is known more formally as the "Last Man in America to Know Who Won the Super Bowl," and its participants call themselves "knowledge runners," as they attempt to escape being informed about who won that year's Super Bowl. It is played on an honor system, and the goal is to go as long as possible without becoming informed of the Super Bowl champion.[45] This is a contagion process that is difficult to evade. First, it begins with a third of the U.S. population being "infected" with the knowledge of who won the Super Bowl, as they watch the game directly. Next, it is very hot news—not only is it a central topic of conversation for several days afterward, but it is also a top story in many news outlets.

Trying to avoid hearing hot news is actually quite a challenge—it requires carefully altering one's habits to avoid a lot of media, conversations, and people. A fascinating aspect of the challenge is that it is nearly impossible for the contestants to last very long. The many ways in which they quickly "die" (learn who won the Super Bowl) are amusing. Contestants last only hours or at most a few days, with only occasional contestants surviving for more than a week. The record reported on the challenge's Web site for shortest time is eight seconds and the longest being an outlier of several years. When a contestant inevitably succumbs, they are supposed to let others know of their cause of "death." The list includes numerous forms of social interaction. A partial list of what they report is "Death by airline stewardess, Death by professor, Death by roommate, Death by college friend, Death by wife's whooping and hollering (just 8 seconds in!), Death by friend at a rest stop, Death by idle conversation, Death by sabotage in AP Biology class, Death by CNBC news meeting, Death by

Black History Month conversation (seriously). The causes of death also include long lists of emails and texts, broadcast, social and other media, and apps.

These lists make it clear how many different types of interactions people have that can convey information, and that people may exchange information that is completely unrelated to the primary purpose of their interaction. This can lead people to have enormous degrees when it comes to learning about very topical information, which means that the network for such diffusion is highly expansive with large basic reproduction numbers and very short distances between people.

Basic reproduction numbers, phase transitions, giant components, and externalities all play prominent roles in many forms of diffusion and contagion, well beyond the spread of disease and news. Some fascinating twists appear when what is spreading is more than a germ, as we will now see with financial contagions.

# 4 · TOO CONNECTED TO FAIL: FINANCIAL NETWORKS

*"But the world is ever more interdependent. Stock markets and economies rise and fall together. Confidence is the key to prosperity. Insecurity spreads like contagion."*

— TONY BLAIR

*"Fear and euphoria are dominant forces, and fear is many multiples the size of euphoria. Bubbles go up very slowly as euphoria builds. Then fear hits, and it comes down very sharply. When I started to look at that, I was sort of intellectually shocked. Contagion is the critical phenomenon which causes the thing to fall apart."*

— ALAN GREENSPAN

Increasingly interconnected economies make it possible for a collapse of real estate prices in Las Vegas to affect financial markets in London and Hong Kong. An investment scandal at a French bank can cause prices of shares of other banks around the world to drop.

Even though financial networks involve a variety of organizations, and their relationships are different from the personal ones we have talked about to this point, they are networks built and occupied by humans, and are full of externalities. Thus, our discussion of contagion provides an excellent point of departure for understanding the spread of financial distress.

However, the reason that financial networks get their own discussion is that there is a countervailing effect that makes the workings of financial contagion subtler than the spread of a disease. Globalization leads not only to more interconnections, but also to more diversified investments and safer portfolios overall.[1] New contacts in a disease network help it spread faster and more widely, while in a financial network new contacts also help spread risk around so that it is better absorbed.

How do these opposing effects play out? Why is it that we still face world recessions and global financial contagions?

Many important financial markets remain remarkably under-diversified for a variety of reasons that we will explore in this chapter. This leads to a dangerous sweet spot for financial contagions in which the network is sufficiently connected to enable widespread contagion, but not so well diversified to prevent one institution's failure from dragging others down with it.

A look at a particular global financial crisis is an illuminating starting point to understanding how a financial cascade can happen.

## Anatomy of a Global Financial Crisis

Lehman Brothers, one of the world's largest investment banks at the time, filed for bankruptcy on September 15, 2008. The drop in the U.S. stock markets that day, of more than 4 percent, was just a preview of what lay ahead. Lehman Brothers' failure was largely due to what can now be seen as excessive exposure to subprime mortgages. They are subprime because the borrower's credit history, and/or the value of the property on which the money is being borrowed, involve significant risks.

Sheila Ramos took out several such mortgages. She had moved to Florida after spending years in Alaska working jobs in beauty salons, a retail chain, and eventually building concrete enclosures for utility companies. Tired of working jobs she did not enjoy, she uprooted and moved south, buying a $300,000 house, free of any debt, using her own savings and money she received from the sale of her parents' home. When she got restless for work and a source of income, she bought a local lawn-care service that she worked herself together with one of her sons. To pay for the business, Sheila took out a $90,000 mortgage on her home. All was fine until an automobile accident sidelined her and her business. Unable to make the payments on her mortgage, Sheila took out a new, larger mortgage for $140,000. She used that money to pay off her previous mortgage, additional rising debts due to her lost income, as well as medical and legal fees, hoping that the remainder would allow her to keep up with her new even larger mortgage payments until she was able to resume work. Unable to resume work, however, and hounded by collectors for her mounting debts, Sheila took out a third mortgage for $262,000 in December of 2006. This would allow her to pay off all of her previous debts, with another cushion that she hoped would tide her over on the new mortgage payments until she was able to work again. Each mortgage involved higher payments, and they were also adjustable and they grew over time.

The mortgages were thrust upon Sheila Ramos by a combination of unfortunate conditions, aggressive sales by the mortgage issuers, as well as decisions that she came to regret. The mortgages also made little sense from the perspective of the companies that were eventually left holding them. Sheila was later to find out that although she told the mortgage broker that she was unable to work and that her business was idle, her application was submitted listing her as employed with a $6,500 monthly income. It is not surprising that her mortgage saga ends sadly, with foreclosure and her moving into a tent with her grandchildren of whom she had custody.[2]

One might take some solace in the eventual bankruptcy of Sheila's mortgage broker's company, but the catastrophe that such loans caused extended well beyond the people directly involved. Her story is extreme, but not atypical, as many people were attracted by the ease of refinancing and obtaining some quick cash against the value of their home. There were also many first-time buyers who found it surprisingly easy to buy a home—from people landing jobs in a then booming Las Vegas to those moving to growing suburbs in New Jersey. Borrowers of all types were aggressively pursued by issuers who were paid a fee for each mortgage, with little risk to themselves given the ease of reselling the mortgages.

The fact that many of the loans had been issued with very creative accounting on the applications was later seen by the courts to be a breach of contract, but even in the most notorious cases, such as that of Countrywide Financial (the largest issuer of mortgages), it was hard to prove outright fraud rather than carelessness and a failure to live up to the terms of a contract. They could not have done it alone. Most subprime mortgages issued in the U.S. passed through two large government-sponsored enterprises: Fannie Mae and Freddie Mac,[3] which held some and repackaged and resold others with some guarantees. Fannie Mae and Freddie Mac would later be sued for their lack of oversight and the resale of securities that were not as advertised. The soaring market was also fueled by some of the lowest interest rates seen in decades and encouragement to grow the number of homeowners from sections of the U.S. government such as the Department of Housing and Urban Development. By 2008, Fannie Mae and Freddie Mac owned or guaranteed more than *five trillion dollars'* worth of mortgages and mortgage-backed securities, nearly half of the enormous market.

The runaway mortgage train derailed in the summer of 2008, but its wheels had already been coming off the tracks for some time. Borrowers saw increases in their rates after the low introductory rates expired, and many like Sheila, who had been encouraged to take mortgages that were beyond their means, began to default. Home prices fell with the ensuing foreclosures, causing more people to be holding mortgages that exceeded the value of their homes. The increasingly obvious inevitable train wreck eventually sounded alarms. By early July 2008, the world was finally worrying about Fannie Mae's and Freddie Mac's enormous exposure to the market. It became all but impossible for them to sell mortgage-backed securities and to do their business as it became clearer that the guarantees Fannie and Freddie had made on many of the mortgages that they resold were essentially worthless, as were the many bonds that they had issued to finance their massive business. Given the trillions of dollars involved, the government had to step in to avoid a complete collapse of the market.

This brings us back to Lehman Brothers. As one of the largest underwriters of mortgage-backed securities, it had been immensely profitable during the run-up prior to 2007. But the dangers of the popping housing bubble became apparent when two of Bears Stearns' hedge funds, also heavily invested in mortgages, failed. Wider-spread fear then led to a major drop in Lehman's stock. Rather than face reality, Lehman doubled down, working more carefully on its accounting than its portfolio. Although it began unwinding some of its positions by the summer of 2008, it was too little too late.

Now we come to the network part of our story, where financial distress spreads beyond those directly involved in the mortgage markets. Given the government's role in avoiding a catastrophe by help-ing Bears Stearns be absorbed by JPMorgan Chase earlier that year, it was rational to expect some support for Lehman Brothers. Unfortu-nately, even though Lehman Brothers was far "too connected to fail," it was left to fail.

A few days after Lehman Brothers' declaration of bankruptcy, AIG (American International Group) had to be bailed out. AIG was the major market maker in credit default swaps, insuring hundreds of billions of dollars of mortgages.[4] The bankruptcy of Lehman Broth-ers sent panic into that market and suddenly it was no longer clear whether AIG could back all of the low-price insurance it had issued presuming that it would never have to pay out on much of it.

The anticipated liquidation of Lehman Brothers' large portfolio of mortgages then led to further panic in the mortgage markets, push-ing more investment banks and investors over the brink. In addi-tion, many hedge funds used Lehman Brothers as a broker and had tens of billions of dollars' worth of their investments flowing through Lehman, which had used those assets as collateral and other means of leveraging its own investments. It also turned out that some prom-inent money market funds, which are thought of as some of the safest short-term investments, were overly exposed to Lehman Brothers. For example, Reserve Primary Fund held more than three quarters of a billion dollars in Lehman Brothers debt. They "broke the buck"—meaning that people who had invested in their money market fund were paid back less than they invested—a worse investment than just sitting on cash.

This spread outward as a loss of faith in money market funds hit companies who depended on those funds for short-term borrowing—companies with no connections to mortgage markets or the firms involved in those markets. Similarly, the interbank lend-ing market—an essential tool that banks use for short-term balanc-ing of their portfolios and deposits—completely dried up, as nobody was any longer sure which banks were solvent and which ones would soon be zombies. For some time thereafter, the Federal Reserve had to step in to fill the hole left in the interbank market.

Within a year and a half, the Dow Jones Industrial Average lost more than half its value: from a high of more than 14,000 in Octo-ber of 2007, it fell below 7,000 by March of 2009. The London-based FTSE 100 would lose more than 40 percent of its value over the same period as banks in the U.K. and Ireland failed. The major stock markets around the world saw similar drops, for instance with the major indices in Hong Kong, Shanghai, Tokyo, Mumbai, and Frankfurt all losing more than half of their value between late 2007 and early 2009. Iceland's financial problems became so dire that its government became insolvent.

Many have argued about whether saving Lehman Brothers would have made a small or large difference in the subsequent crisis. The market was very interconnected and many financial institutions were overly exposed, directly or indirectly, to the disaster in the subprime market. Although some of the ramifications of Lehman's failure may have happened eventually in any case, the panic it caused in many markets led the impact to be much wider than the financial markets in which it originated. From a network perspective, stopping contagions at earlier points is always easier and cheaper than letting them play out and then trying to clean things up afterward.

There are three main reasons that the U.S. government, in the form of the U.S. Treasury and the Federal Reserve, did not intervene.

First, some officials at the U.S. Treasury wanted to send a message that large private companies could not all expect bailouts. The Treasury had already been involved in bailouts of Bear Stearns, Fannie Mae, and Freddie Mac, and many people, including congressmen, expressed a view that enough was enough. However, as Darrell Duffie (a professor of finance at Stanford University, and a leading expert on financial markets, who also happens to have been my dissertation advisor) said, "When the house is burning down and the fire crew is there and the firefighters are hosing down the house putting the fire out, it's not the time to turn off the hose and give the homeowners a lecture about smoking in bed."

Second, critical people did not have the information they needed to predict the consequences of letting Lehman fail. The minutes from the Federal Reserve Board meetings at the time reveal little awareness of what would ensue. A detailed network picture of the financial system was not available then, and is remarkably opaque even today. If various arms of government (including Congress) had had a clearer network picture of the exposures at the time, they would have—or at least should have—acted very differently.

Third, there were also questions about how the Federal Reserve could have best intervened under the legal restrictions on its activities. This was uncharted territory and it was not clear what the Fed was really entitled to do.

The consequences of not intervening appeared quickly and were eventually widespread and catastrophic. Letting Lehman Brothers fail, in hindsight, was a huge blunder and ended up costing much more in the subsequent bailouts. Eventually the trillions of dollars spent by governments around the world did manage to stop the contagion from truly playing out like a plague, but it did not save the world from long and painful recessions. Even so, we were lucky not to see the full cascade of failures that could have ensued. What if the U.S. Treasury had not intervened on behalf of Fannie Mae and Freddie Mac, and if there had not been a bailout of AIG, and if various governments around the world had not propped up their banks and a variety of businesses?

Clearly lessons had been learned from the Great Depression, when it took years for the government to take the right actions after large-scale bankruptcies of the many banks and others who had been over-extended in a soaring stock market (instead of a soaring subprime mortgage market). The ensuing sell-offs, panic, and freezing of capital markets caused huge contractions in investment, business, and ultimately wages and consumer spending, sending the world economy into a deep spiral. Even then the network effects were apparent. Just as one example, the crash of Wall Street cut short the large lending to Germany that was enabling it to make its payment of World War I reparations. As Germany halted its payments, panic ensued and investors stopped lending there, leading to widespread business collapse and a record-breaking depression, furthering the political and economic chaos during which the Nazi party gained a solid toehold. Of course it is impossible to know whether World War II would have been avoided had the spread of the Great Depression been cut at an early stage, but the financial contagion of the Great Depression is unmistakable.

A strong contrast between the international trade network of then and now is that the network was much more fragmented at the time of the Great Depression. For instance, China, Japan, and the Soviet Union had limited business with the West in the 1930s, and their banking systems were almost entirely insulated, and thus they saw almost no effects from the complete economic meltdown that was hitting parts of the Americas and Europe. In fact, it was a period of large growth in industry for the Soviet Union.

## What Is Special About Financial Contagions?

> *"So, first of all, let me assert my firm belief that the only thing we have to fear is fear itself—nameless, unreasoning, unjustified terror which paralyzes needed efforts to convert retreat into advance."*

—FRANKLIN D. ROOSEVELT,
FIRST INAUGURAL ADDRESS, MARCH 4, 1933

Our basic network analysis provides insights into economic contagion. Centrality in a financial network can help identify who is too connected to fail and it is becoming a primary tool in assessing risk. The connectedness of the network serves as a first assessment of the risk of potential contagions in the global network.

However, there are fascinating twists.

First is that financial markets involve a multitude of different players and types of interactions. Whereas a flu makes its way from one person to another by the spread of a virus, a financial insolvency can make its way from borrowers to banks to insurers, to the many others who do business with them, to markets, and then to employees and shareholders throughout an economy. The connections in the network involve all sorts of transactions and contracts—from loans to partnerships to insurance contracts to elaborate securities to the simple sales of goods and assets—with the key being that something of value is owed from one party to another, and so the wealth of one party depends on the other's. This enormous scope of the network makes it hard for investors and governments to truly assess the rare but potentially catastrophic risk that underlies financial contagions.

A second twist is that having more connections does not always translate into a greater danger of "infection" in financial networks. Having many counterparties can be safer than contracting with just one. This is one of the most basic principles of investing: diversifying a portfolio lowers risk. If exposure is spread among enough counter-parties, then a default by any one of them becomes inconsequential. This makes financial contagion fundamentally different from other forms of contagion. Having sex with many partners increases one's risk for catching sexually transmitted diseases greatly compared to having sex many times with the same partner. In contrast, by spreading business across more relationships, companies can become less exposed to the shocks that might hit any particular region, market, or supplier.

This leads to an interesting trade-off in economic networks.[5] Initially, as companies connect to each other via trades, securities, and various contracts and obligations, the threat of financial contagion increases because the enterprises become interdependent—a network takes shape and we grow above a basic reproduction number of one. For instance, if each organization has two or three major counterparties, then the network can be quite connected in terms of having a giant component. Moreover, with just a few major counter-parties each organization is at substantial risk if one of its partners defaults on an obligation or is unable to deliver on a promise. However, as we continue to add connections, the level of systemic risk eventually falls as each given institution becomes more diversified in its exposure and less prone to fail as a consequence of any one of its neighbors failing. Even though the network is becoming more densely connected, the chance for cascades along any of those connections is dropping.

The worst case for financial contagion is the middle ground. There are enough connections to generate a connected network in which companies are all indirectly connected to one another. However, there are still few enough connections so that most companies do a large amount of their business with just a few others and could be dragged down by the insolvency of a business partner.

This trade-off is illustrated in Figure 4.1. For instance, think of each node as a bank that has some investments on its own, and the links as representing investments and contracts with other banks. Those relationships might involve the purchase of a share of the investments of the other bank (e.g., mortgages) or bonds issued by the other bank, or short-term loans, et cetera, all of which are claims on the assets and investments of the counterparties. The thickness of the relation-

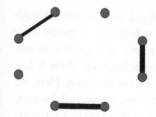

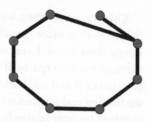

a) Relatively safe: Modularity means contagion cannot spread. However, the lack of interaction means that potential gains from trade may not be realized, and individual banks may be at greater risk from direct failure.

(b) Risky: Contagion can reach widely and each bank has few counterparties and substantial exposure to its counterparties.

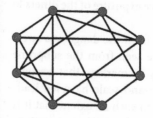

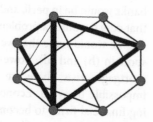

elatively safe: Although contagion could spread, no nk has large exposure to any of its counterparties.

(d) Risky: A core of four banks are very exposed to each other and the density of the overall network means that if the core became insolvent it could spread.

**Figure 4.1: Four financial networks that differ in the likelihood of a financial contagion. The thickness of the links indicates the level of interdependency between two banks.**

ships indicates the fraction of the banks' investments that are with each other bank. I am abstracting from all sorts of details and just illustrating the basic structural dependencies.

As we see in panel (a), a very disconnected network is relatively safe from having widespread contagion. But this comes at a large day-to-day cost: there is not much trading between the different financial institutions. The lack of interaction means that the banks are not sharing the idiosyncratic risks that they face, and so their returns are unnecessarily noisy and risky.[6]

As we move to panel (b), the network becomes connected. This is a particularly dangerous case as it has moved to a situation in which there is a path of potential contagion from every institution to every other one, *and* each bank has substantial exposure to its counterparties and is still not well-diversified.

Risks improve when we move to panel (c), as now even though the network is more densely connected, the banks have spread their interactions around and so lowered their level of exposure to any particular counterparty. If one bank becomes insolvent there is less of a chance that it will infect and drag down a counterparty. Here, we see the first real difference between financial contagions and diseases, as things become relatively safer moving from (b) to (c) in a financial setting, while it would be the reverse in the case of a disease.[7]

Finally, when we move to (d) things have gotten worse again, since we have increased the exposure of four of the banks to each other. If one of them were to become insolvent then the core of the four major banks would be in peril, and the cumulative exposure of the others to that core becomes a problem.

One would hope that we would be far from the dangerous sweet spots in (b) and (d); however, as we have seen from the subprime mortgage crisis, there are major players in markets that become impossible to avoid. Economies of scope and scale have led leading financial firms to become enormous, to such an extent that it is impossible for most other firms not to have extensive transactions with the largest players.

A third twist with financial markets is that "infection" can occur even without any contact: a bank might become insolvent even without having any of its investments fail. Fear and uncertainty can be as damaging to financial markets as the cascading failures due to some bad investments. If we all woke up tomorrow and believed that Bank X would be insolvent, then it would become insolvent. In fact, it would be enough for us to fear that others believed that Bank X was going to fail, or just to fear our collective fear! We might all even know that Bank X was well-managed with healthy investments, but if we expected others to pull their money out, then we would fear being the last to pull our money out. Financial distress can be self-fulfilling and is a particularly troublesome aspect of financial markets. Such fears underlay some bank runs.[8]

This distinguishes financial markets from other markets. If you are about to buy some apples from your local grocer and fear that the grocer might go bankrupt, it does not affect your decision to buy those apples. You might have to go elsewhere for your apples in the future, but that fear would not discourage you from your purchase. If you are about to deposit money in a bank and fear that the bank's business may soon be disrupted, it makes a huge difference. Perhaps you just hear a rumor—you don't have to believe the rumor or even believe that others will believe the rumor—all you have to believe is that the rumor may cause others to withdraw their money from the bank. Once people lose confidence in the financial system in general, savings and investments really are at risk. The hoarding of cash during the Great Depression was a serious problem.

This is an ageless aspect of finance, and of investments more generally, as we see in Roosevelt's quote from the Great Depression. The loss of more than 50 percent in world stock markets between late 2007 and early 2009, and then the subsequent rebound a few years later, cannot be rationalized as a huge loss in the actual values of the companies and then a miraculous rebound; instead it was uncertainty about which companies might end up insolvent and how wide and deep the recession would be and the fear that accompanied that uncertainty. The fact that markets eventually regained their footing does not mean that all ended well. There was a substantial loss of employment, production, and consumption during the downturn, and the disruption of basic investments and economic activity lasted for some years. Uncertainty in and of itself is costly and debilitating.[9]

This third twist was on John Maynard Keynes's mind when he discussed what has become known as the Keynesian Beauty Contest in his *General Theory of Employment, Interest and Money*. He described a contest from a newspaper in which people try to pick the six "prettiest" faces out of a set of a hundred pictures. The twist was similar to that in our discussion above. "Prettiest" was defined to be the faces picked by the most people. As Keynes states (page 156), "It is not a case of choosing those which, to the best of one's judgment, are really the prettiest, nor even those which average opinion genuinely thinks the prettiest. We have reached the third degree where we devote our intelligences to anticipating what average opinion expects the average opinion to be. And there are some, I believe, who practise the fourth, fifth and higher degrees."

This twist means that some aspects of investing are uncoupled from the values of the underlying assets, allowing bubbles in share prices to exist as well as fear-induced runs. At times, investing is as much about predicting what others will be willing to pay, and for how long, as it is about understanding the true value of an investment.[10] Networks and perceptions enter again—since prominent individu-

als can have a disproportionate impact on the value of a stock—by
expressing an opinion or starting a rumor, especially if we expect
others to be paying attention to the same individual.[11]

## Free Markets and Externalities

*"The very nature of finance is that it cannot be profitable unless it is
significantly leveraged . . . and as long as there is debt, there can be
failure and contagion."*

—ALAN GREENSPAN

Adam Smith was rightly impressed by the workings of markets when
he coined the term "the invisible hand." There are many markets that
work well when completely free and open, unfettered by intervention. Many goods and services that people consume, from bread to
haircuts, involve at most minimal externalities so that the individual
costs and benefits to producing and consuming such goods reflect
the societal costs and benefits. Also, economies of scale in many markets are small enough so that many firms can compete, with none
becoming so large as to have a substantial impact. Unfortunately,
neither of these is true of financial markets: they involve substantial
externalities and economies of scale.

The externalities in financial markets come in the form of the network of potential consequences from one firm's mistakes. One firm's
insolvency and default on payments can lead its counterparties to
become insolvent, causing cascades of failures and associated costs of
bankruptcy. Those costs are large. In a review of bankruptcy costs,[12]
Ben Branch, an economist at the University of Massachusetts-
Amherst, estimates that claimants on a firm typically get 56 percent
of the (book) value of the bankrupt firm before it became insolvent.[13]
Typical recovery rates on bonds are in the 40 to 50 percent range,
and even if the debt is secured or has priority in bankruptcy that
still might only be 70 percent. The people who are owed money by

the bankrupt firm are losing a large fraction of their investment, and losses of this magnitude can lead them to become distressed too.

These externalities mean that incentives are not aligned: Lehman Brothers not only continued to underestimate the risks of a heavy exposure to subprime mortgages even as late as early 2008, but they were also mainly worried about their own profits, not the catastrophe that they would cause if they were to go bankrupt. Certainly nobody tries to go bankrupt, but the full risk to society of a firm's investments can be many times larger than the direct risk to the firm, especially given the enormous leverage that is common in such markets. Many investments are made with borrowed money, or a variety of other arrangements, which only exacerbates the problem.

To illustrate the point, if you are employed by a company that is having cash problems and starts delaying your paychecks, and eventually stops making those payments, then you may start missing mortgage and/or car loan payments. If that leads to your being called into default on those payments, it leads to serious costs to you. The same is true when one company becomes insolvent and another company to whom it owes money ends up getting only a fraction of what it is owed, often with substantial delay. That next company can then have difficulties paying what it owes, which is compounded by further legal costs and losses, and can lead it to delaying or stopping payments, and this then cascades, with more costs accumulating at each step.

This would not be too big of a problem if the networks were well-diversified and each firm had many counterparties, none having too large an exposure. In that case, any single firm's insolvency becomes a small blip that is easily absorbed by its many partners with little lasting impact. However, this is where the economies of scale cause problems.

In financial markets there are substantial advantages to being huge, and so there are key firms that are both too big to fail and too connected to fail. For instance, the size of Fannie Mae and Freddie Mac made them hard to avoid in the subprime mortgage market: many large firms were owed huge amounts of money by Fannie Mae and Freddie Mac. Fannie's and Freddie's failures would not have been easily absorbed blips: the government had no choice but to step in when they became insolvent—eventually taking them over and swallowing enormous losses. Had it instead let that snowball keep running downhill, it would have grown in size and could have cost even more trillions of dollars in costs to the economy before it came to rest.

As evidence of "bigger is better" pressures in financial markets, the number of banks continues to drop, and the size of the largest banks continues to grow. In 1980 there were over 14,000 commercial banks recognized by the FDIC (Federal Deposit Insurance Corporation) in the U.S., yet by 2016 that number was just over 5,000. This consolidation was not due to a contracting industry: the amount of FDIC-reported assets of these institutions went up by a factor of more than eight, moving from under $2 trillion in 1980 to over $15 trillion by 2014.[14]

The consolidation of financial services into enormous banks is not just happening within a narrow part of the industry, but also cutting throughout the industry as a whole. The Glass-Steagall Act of 1933 had separated investment banking (underwriting the issuance of securities, market making in various securities, and handling mergers and acquisitions, among other activities), commercial banking (taking deposits and issuing loans), and insurance (issuing insurance policies). The Gramm-Leach-Bliley Act (the Financial Services Modernization Act) of 1999 undid that. John Dingell (a congressman from Michigan), in a prescient debate, predicted that it would lead to bank holding companies becoming too big to fail. Individuals could now hold savings and investment accounts at the same institutions. It can be convenient to have one-stop financial shopping. However, the resulting behemoth financial institutions are far from ideal for reasons well beyond their too big to fail sizes, as they can be betting against the trades of their depositors and have basic conflicts of interest when advising and managing money.[15]

Most important, the capital is not spread evenly among banks; much of it is concentrated at the very top. The ten largest banks in the world—the top four of which are in China—had assets of almost $26 trillion in 2016. To put that in perspective, the GDP (a measure of the total value of production) of China and the U.S. combined was just over $29 trillion in 2016 and the world total GDP was just over $75 trillion. In 1990 the five largest banks in the U.S. accounted for roughly 10 percent of the total assets in domestic banking industry but by 2015 they accounted for almost *45 percent*. Total assets actually went *up* substantially after the subprime mortgage crisis, as the five largest banks controlled just above 35 percent of industry assets in 2007. It is almost impossible for many companies to avoid having substantial dealings with one of the largest banks in the world.

Figure 4.2 gives an impression of what the network of the largest of the U.S. banks looked like even before this further consolidation. It shows the sixty-six largest banks in what is known as the Fedwire system, which is a way in which banks and other related financial institutions transfer funds from one to another. This is at a very aggregated level, so it abstracts from the details of what is actually being transacted between the banks. Nonetheless, it gives us a high-level impression of who is doing how much business with whom, and thus who might experience serious disruptions in its business if one of its counterparties were to become insolvent.

We see several telling patterns. The three or four largest players have enormous size and large volumes of interactions with each other. Also, the twenty-five main banks form a tightly clustered core, and then each of the peripheral banks (often regional banks) feed

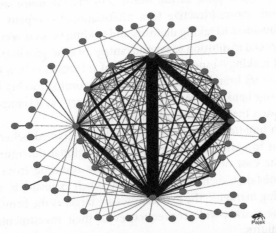

Figure 4.2: The sixty-six largest banks in the Fedwire system (accounting for three fourths of the total volume). The data are from 2004, before the subprime mortgage crisis and when concentration in the banking industry was much lower than today. The twenty-five most active banks form a densely connected cluster. The link width corresponds to the amount of transactions between two banks. Reprinted from *Physica A: Statistical Mechanics and Its Applications*, Vol. 379, Kimmo Soramäki, Morten Bech, Jeffrey Arnold, Robert Glass, and Walter Beyeler, 317–333. Copyright (2007), with permission from Elsevier.

into one or two of these core players. This quite distinctive pattern is called a "core-periphery network," and is the shape of many financial networks.[16] It becomes easy to see why we should worry about the failure of one of the largest core banks, as it could affect the other large banks and the core, and ultimately the periphery. This is not so unlike panel (d) from Figure 4.1!

What underlies the enormous size of the largest banks and their continued growth? There are many features that lead bigger to be better in financial markets. On a very basic level, one can save on fixed costs of information and communication technologies. There are many costs associated with building and maintaining databases and accounting systems that do not depend on how many entries they hold. By merging, two banks have to pay those costs only once instead of twice. On the investments and business side, having more money to invest allows one to build more diversified portfolios and also opens up opportunities to invest in large projects that are beyond the reach of small investors. It also permits one to diversify over broader geographic regions and branches of business. Having a presence in many parts of the world and offering more services makes a bank more attractive to a multinational company. Also, expanding business to taking deposits and trading various securities allows a financial institution to take advantage of synergies in trading and market making—especially in terms of information. On top of all of this, comes branding and reputation: becoming too big to fail makes the large bank a more attractive counterparty, as it is implicitly insured in ways that smaller banks are not.

Although building these larger and better-diversified behemoths allows them to easily absorb any small or regional downturns, it comes at the cost of the enormous disasters resulting from their inevitable mistakes, such as having far too much exposure to a sub-prime lending market.[17] The externalities are large, as the firms don't consider the market-wide consequences of their investments and potential failures.

In addition to these externalities, coupled with size of the main players, there are pressures that push companies to do business with just a few counterparties. The costs of contracting and doing business can make it much more attractive to contract with one or two counterparties rather than spreading business widely. This is easy to see in our own lives: maintaining multiple bank accounts becomes costly and time-consuming. We don't worry too much since our governments implicitly or explicitly insure many accounts—and the same is true for many businesses. We see that quite clearly by looking at the peripheral banks in Figure 4.2: they do most of their transactions with one, two, or three other banks. Not only is it costly to maintain many contacts, but also doing more business with just one counterparty leads to larger accounts, which may get better treatment and attention. Preferred customers exist at all levels of business. The larger interdependence can help two parties work out a contract when unforeseen issues arise.[18] Larger interdependencies also lead to preferential treatment in turbulent times, and the repeated interaction with a partner can help build trust that is reinforced by the value of the future relationship. A company is less tempted to gouge a partner with whom it expects a long and fruitful future business stream.

There is more than one aspect of a large core bank's decisions that can hurt others. One is that it chooses overly risky investments and becomes insolvent. The second is that it does too much of its business with a few partners and thus leaves itself (and the rest of the network indirectly) exposed to one of its large partners going bankrupt. These are worth distinguishing: the first externality is that a bank might put itself at more direct risk of bankruptcy and starting a contagion than the rest of the network would like, while the second externality is that it positions itself in the network in a way that makes it more likely to become infected by others and then to perpetuate a contagion.

Financial networks are also now very international. Figure 4.3 shows us the debt obligations from one country held by banks within another country, which comes from research I did together with economists Matt Elliott and Ben Golub, two of my former students. It focuses on just a few of the key European countries in the Greek debt crisis. As an example, almost $330 billion worth of Italian debt was held by French banks at the end of 2011. This is reflected in the large arrow from Italy to France.[19] French banks then had substantial exposure to a drop in the valuation of Italian debt, which, for instance, can be influenced by the health of Italian banks. The Italian banks hold a lot of German debt, and German banks hold a lot of Greek debt, and so forth.

Even though Greece is a relatively small country, the exposures in question were large enough to be a big problem. A few percentage

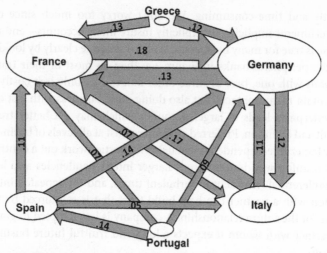

Figure 4.3: The arrows show how much sovereign debt (government-issued bonds) of each country is held in each other country based on data from the end of 2011. The widths of the arrows indicate the size of the exposure (excluding dependencies of less than 5 percent). For instance, 13 percent of Greek debt at the end of 2011 was held by French banks. Figure and data from Elliott, Golub, and Jackson (2014).

points change in returns can move a financial institution from being profitable to not, and these holdings are often heavily concentrated. There was good reason for the troika of the IMF, the European Central Bank, and a fund from European countries to bail out Greece via emergency loans and by buying hundreds of billions of euros' worth of Greek bonds to keep their value from collapsing and taking many banks with them.

To summarize, a variety of forces push financial networks to the sweet spot that enables contagion:

- Economies of scale and scope give bigger banks many advantages over smaller ones and so a substantial fraction of business is concentrated among the largest banks.
- Banks and other financial institutions have reasons to trade with each other, leading to a connected network.
- Having larger dealings with a few counterparties, rather than spreading that business widely, saves relationship costs and can lead to preferred treatment. Thus, there are critical connections

in financial markets that involve substantial exposure of one company to another.

- The substantial costs of bankruptcies can cause one company's insolvency to drag others down with it.
- Externalities on counterparties and potential cascades are not taken into account by companies when choosing their portfolios and partnerships.
- Uncertainty and fear cause investors to pull money and stop lending, which can quickly push both distressed and otherwise healthy financial institutions into insolvency.

These forces—especially in combination—mean that financial markets require oversight and care.

## Regulation

*"But the position is serious when enterprise becomes the bubble on a whirlpool of speculation. When the capital development of a country becomes a by-product of the activities of a casino, the job is likely to be ill-done."*

—JOHN MAYNARD KEYNES, *THE GENERAL THEORY OF EMPLOYMENT, INTEREST AND MONEY*

How should we regulate financial markets? The divergence of views on this question is enormous. Some people simply argue that financial markets work well completely on their own. Such people are either ignorant of the size issues, externalities, and network problems present, or simply choose to ignore the implications because they have a stake in the game. They either stopped learning about economics after the first lesson on markets and never learned about externalities; or they stand to profit from their position in the network. Others echo a range of economists, from Alfred Marshall, to Ludwig von Mises, to Milton Friedman, who—either from a distrust of governments, a strong philosophical belief in the individual, or a faith that competitive forces can overcome anything in the long run—conclude that we are better off minimizing government intervention or avoiding it altogether.

Governments have shortcomings, some much worse than others. Nonetheless, the largest enterprises are so huge that governments have repeatedly been forced to step in and bail them out throughout the world's financial history, and so government intervention is inevitable. It is then better to have even minimal oversight that avoids disasters before they happen rather than paying to clean them up afterward. Making sure that such large enterprises are not overly exposed to any particular investment or counterparty is an obvious start—part of the thinking behind stress tests. Such oversight is not easy, as the tendency of firms to be secretive about their investments and trading strategies, coupled with the increasing complexity of securities and derivatives, makes it a challenge to get a detailed view of exposures.

Moreover, this is now a global challenge: a default by a U.S. investment bank is not a regional issue, nor is a default on Greek bonds, nor a drop in the Thai currency,[20] nor a change in Chinese housing prices. Tracking the global financial network in fine detail is impossible at present, and so one is left with partial pictures.[21] Making progress will require more careful and unified accounting standards, communication between various governments and agencies, and methods of flagging and defusing dangerous exposures when they emerge instead of waiting until they become obvious because of impending catastrophe. And all of this has to be done while balancing the costs it imposes, both on government and the market participants. It is a bit like brain surgery: it is delicate and we still don't have a fully comprehensive understanding and map of the complex system we are dealing with, but there are clear situations in which it is needed.

It is worth adding that there are many ways in which incentives are misaligned in such complex markets. As we saw with subprime lending, the fact that mortgages were being resold—and not closely inspected by the rating agencies, insurers, and others who should have been doing due diligence—meant that those issuing them profited by issuing mortgages as fast as they could with little attention to whether the mortgage was right for the borrower or would eventually be repaid. This is just one example of what is known as playing with "other people's money"—the person making decisions on the investments is not at risk for the full amount of the investments. Most financial investments are heavily leveraged: much of the money involved is borrowed or from other partners who are not directly involved in the day-to-day choices being made. Traders for large investment banks, mutual funds, hedge funds, and others are rewarded if risks pay off, but face limited losses if the investments go bust. This distorts their decisions, and can lead them to take on too much risk.[22] The implicit bailouts that wait in the wings for "too big to fail" enterprises, mean that they are playing with other people's money—the taxpayers' in addition to any other counterparties whose money they are managing or have borrowed. This exacerbates the misalignment of their interests due to the externalities of broader cascades that exist due to the network structure and costs of bankruptcy.

Governments are thus faced with economies that are based on an enormous and complex web of interdependencies and exposures that is constantly evolving and hard to monitor. Governments are also faced with behemoth enterprises that they explicitly and implicitly insure, who exude externalities, and who have incentives that are not aligned with a country's long-run prosperity. These challenges are compounded by the fact that there are significant economies of scale and scope, and one does not want to simply give up on the substantial cost savings that are associated with large businesses that take advantage of synergies across geographic and business domains.

Regulating markets in terms of which institutions can make which sorts of investments faces its own difficulties. Even when one regulates one market, investors move their money to other institutions that are not covered by the regulation. For example, after the Great Depression banks were prohibited from providing interest on checking accounts. Accounts by savings and loans and investment banks, however, were not regulated and could pay interest. Thus, savings and loans and other financial institutions began to offer checking accounts and easy movement of funds together with interest payments. Not surprisingly, people started moving their money to interest-paying accounts, driving the growth of savings and loans and money market accounts that lay outside the regulations.[23] The competition to attract customers led savings and loans to make shaky investments in order to be able to pay the increasing interest rates that they were promising their customers. This eventually led to more than one thousand savings and loans (out of just over three thousand in the U.S.) to go bankrupt between the late 1980s and 1990s. In addition to the financial stress it caused, this drove people to move their money out of savings and loans and helped fuel the growth of investment banks and a variety of other funds and enterprises. Many of these players were eventually caught up in the subprime crisis by 2008.

Thus, financial regulation has to deal with moving targets that continuously adjust and evolve to circumvent restrictions.[24] This is something that few governments are currently equipped to do. Developing better systems for comprehending the network of contracts and exposures is an important starting point.

## *Popcorn or Dominoes?*

There is a key distinction regarding which crises are really networked: is a financial crisis a case of "popcorn or dominoes"? This is a metaphor of Eddie Lazear (a coauthor and friend), who was the chairman of the Council of Economic Advisors from 2006 to 2009 at the time of the subprime crisis.

Network contagion is a dominoes story: one default causes problems for others, like rows of dominoes knocking each other over. Another feature of some financial crises is like popcorn popping. Popcorn kernels that are all bubbling in the same hot oil eventually come to explode around the same time for the same reason; and one might mistakenly believe that they are causing each other to pop.

The subprime crisis involved both popcorn and dominoes. The hot oil that many companies were boiling in was clearly the deteriorating housing market and defaults on mortgages.[25] Even though it can be difficult to pinpoint which companies would definitely have gone bankrupt completely on their own, it is not hard to see that government takeovers and bailouts, especially of Fannie Mae, Freddie Mac, and AIG, were essential—a default on the trillions of dollars of contracts that they had with other firms in the U.S. and around the world would have been catastrophic.[26] Moreover, there were many large banks and other companies that were the first row of dominoes, which were not directly at risk due to failing mortgages but were indirectly exposed via their large interactions with the companies that were popping.

The "Financial Crisis Inquiry Report"[27] (commissioned by an act of Congress) makes it clear how the impending dominoes spurred the government's intervention in the propping up of AIG, the takeover of Fannie Mae and Freddie Mac, and the bailouts and subsidized sales of various institutions. For example, they quote an email (page 346) in September of 2008 from Hayley Boesky of the New York Fed to William Dudley, who was head of the capital markets group at the New York Fed: "More panic from [hedge funds]. Now focus is on AIG. I am hearing worse than LEH [Lehman Brothers]. Every bank and dealer has exposure to them." As the report states (page 347), "AIG's bankruptcy would also affect other companies because of its 'non-trivial exotic derivatives book,' a $2.7 trillion over-the-counter derivatives portfolio of which $1 trillion was concentrated in 12 large counterparties." This is consistent with the who's who list of banks that would eventually get government payments of billions of dollars to cover payments they should have gotten from AIG.[28] The report concludes (page 352): "Without the bailout, AIG's default and collapse could have brought down its counterparties, causing cascading losses and collapses throughout the financial system."[29] The reference to dominoes could not be clearer.

## Takeaways: Flying Jets Without Instruments

With increasing globalization, financial networks are becoming more connected and the most central players are growing larger than ever. Although the connectedness of the networks and size of many of the most central players helps the system easily absorb normal and even some large losses, the potential for worldwide recessions caused by unexpectedly large shocks to key nodes has grown.

Although our network perspective is vital, it does not settle a debate on how to best manage the many trade-offs associated with economies of scope and scale, the increasingly intertwined global markets, and the many externalities and distorted behaviors involved.

Striking the right balance between free markets and regulation is a challenge.

Nonetheless, a network perspective makes clear that there are enormous externalities, and that in order to assess systemic risk, we must have a more complete picture of the financial network of exposures: real dangers are completely missed by looking at financial institutions one at a time. Diffusion centrality-like calculations for financial markets are feasible with the right information. Central banks, and other national and international government branches and agencies, not to mention financial institutions themselves, are essentially flying jets without instruments. They are making rapid decisions that steer complex machinery based on limited information. Having more extensive and detailed maps of financial networks would begin to provide the actors, both private and public, with some of the instruments they need to avoid future crises.

More generally, our look at financial networks reemphasizes several points. Networks enable one individual's or institution's actions to affect many others. Such externalities are important in understanding why people's incentives to vaccinate are too weak, and their incentives to take on risky investments are too strong. In addition, there are externalities in the partnerships that people form—having a sexual partner who is not monogamous or a financial partner that has substantial exposure to a risky counterparty are both costly. Finally, people react to what happens in their networks, from changing their travel patterns to turtling up and pulling their money out of markets, with potentially drastic consequences.

## The Caste System

*"Segregation is that which is forced upon an inferior by a superior.
Separation is done voluntarily by two equals."*

—MALCOLM X

The caste system in India permeates the life of its citizens, guiding whom they marry, which professions they undertake, which god(s) they worship, and with whom they interact on a daily basis. The caste system is complex even to those living within it. It has strong roots in Hinduism, but also involves hundreds of millions of Sikhs, Jains, Muslims, and Christians. It evolved over thousands of years, in a society of local tribes and clans, amidst periods of sweeping changes in the ruling classes. India is so large and diverse that the caste system morphs according to local customs and traditions. Despite efforts on the part of British magistrates to formalize a simple caste hierarchy under their rule in the nineteenth century, the system remains complex and nuanced. Even though it defies description, I will try to give an idea of some of its basic structure.

The caste system can be thought of as having four primary groupings—known as "varnas" or "colors"—which have hierarchical connotations: Brahmins (priests and teachers), Kshatriyas (warriors and rulers), Vaisyas (farmers, merchants, artisans), and Shudras (laborers and peasants). There are also "outcasts" and "untouchable" groups outside of the four varnas that include what are now often known as the Dalits ("other laborers") and the Adivasis (various tribal and aboriginal groups). Attached to this is the idea that souls are infinitely lived while bodies and personalities are temporary. Your "karma" depends on the rightness or wrongness of your beliefs and behaviors and can accumulate over lifetimes. Good karma can lead to rebirth in a higher caste. Ultimately, if you come to know your soul then you can achieve "moksha": a divine state of well-being that frees you from the endless cycle of rebirth. The idea that your caste reflects your past karma leads to a perception that your caste is somehow deserved and that you should not aspire to caste mobility, at least not within this lifetime.

Subdividing the varnas are thousands of "jatis" or "subcastes," which more finely align with birthplace, hereditary profession, marriage rules, and religion, and make the system even more constraining in terms of advancement or interaction across groups.[1] Understanding the caste system is further complicated by the fact that although caste correlates with things like wealth and education, there are occasional anomalies, such as untouchables becoming famous politicians or businessmen, as well as Brahmins living in abject poverty.

The rigid limits on opportunity and social mobility imposed by the caste system have led to extensive government programs to lessen its impact. Some of the castes and subcastes are considered disadvantaged and recognized for affirmative action by India's constitution as well as various other laws and programs. Those castes are primarily what are known as the Scheduled Castes and Scheduled Tribes, including the untouchable groups outside of the main varnas as well as a multiplicity of aboriginal groups. Because the caste system is anchored in religious beliefs, loosening its grip on society is a challenge. For instance, a system of reservations provides a minimum number of slots in universities, politics, and government for members of the Scheduled Castes and Scheduled Tribes. This system is loved and hated depending on one's caste and religious beliefs.

In a country in which more than a fifth of the population lives below a very low poverty line, many people outside of the Scheduled Castes and Scheduled Tribes are also severely disadvantaged. More than a third of India's population falls into a category called Other Backward Castes, who in some regions and circumstances are also recognized for affirmative action. All others are often referred to as the Forward Castes (or General Castes or General Merit castes).

Although the caste system defies simple description, its consequences do not. Each caste is a minority of the population, so if caste did not matter most people would marry outside of their own. A recent comprehensive study of cross-caste marriages, however, found that only 5 percent of Indians marry outside of their caste,[2] despite government subsidies for inter-caste marriages. In fact, more than two thirds of the women in the study had met their spouses for the first time on their wedding day. Marriages are still largely arranged relationships, respecting rigid community norms, and often are as much about the extended families as the husband and wife.

The impact of caste goes far beyond marriages. You can see the strength of the divisions created by the caste system in Figure 5.1. Pictured is a network of favor exchange among the households in one of the villages from our microfinance study discussed in Chapter 2. Here nodes are households, and two of them are connected if they borrow from and/or lend kerosene and rice to one another. In such a village kerosene is the main fuel used for cooking and heating, and rice is a primary source of sustenance. In many ways, this network is really the backbone network of the village: households that share kerosene and rice often also lend money, provide advice, and give medical help to one another.

In Figure 5.1, the round solid nodes are the households that lie within the Scheduled Castes and Scheduled Tribes, while the square checkered nodes are those belonging to the Other Backward and General Castes. Even in such a small village, people are *fifteen times* more likely to exchange favors within caste groupings than across them, despite the substantial gains to be had from cross-caste diversification.

Figure 5.1 also makes clear that visualizing networks reveals things that otherwise might escape our notice.

This network reveals a clear additional division within this Indian village, indicated in Figure 5.2. There are groups within the Scheduled Castes and Scheduled Tribes that do not interact—there is just one connection between them. These reflect further divisions. Some subcastes interact almost seamlessly and others not at all—and it is not always easy to predict which will without being part of the local culture.

These sharp divides have profound consequences. Villagers experience health problems, crop failures, wildly changing employment, in addition to a variety of needs for cash, such as dowries. They don't have insurance or much, if any, savings: they rely on each other for help. It is easy to see that such divides become a problem when one considers risk sharing. In a drought, farmers all suffer and cannot

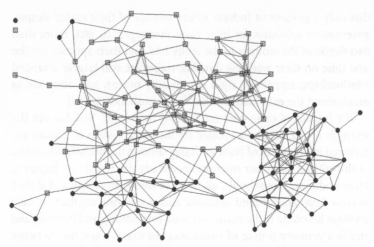

Figure 5.1: Divisions within a rural Indian village. Nodes are households and a link indicates that the pair of households report that at least one borrows kerosene and rice from the other. The round (solid fill) nodes are households that fall within the Scheduled Castes and Scheduled Tribes: those especially considered for affirmative action by the Indian government, and the square (checkered fill) nodes are the Other Backward and General Castes. The positioning of nodes is by a spring algorithm that groups nodes more closely together when they are linked to each other (and is not based on geography, caste, or other household attributes). The frequency of links among pairs of households when both are within the same caste grouping is fifteen times higher than when they are in different caste groupings.

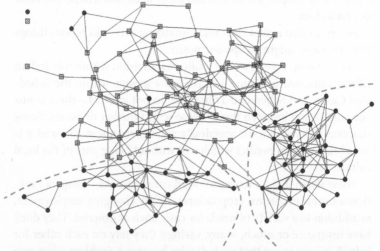

Figure 5.2: The network pictured here has obvious additional divisions that are not captured by the broad caste split, and that are indicated here with the dashed curves.

help each other. If they interact only with other farmers, then the farmers in a village can be starving while the artisans may have more work than they can handle, or vice versa. It has been estimated that between 87 to 90 percent of the money that people gave or lent to others in some villages like these went to someone within the same caste, and that risks were far from being shared well across castes, mirroring the network pictured in Figure 5.2.[3]

One reason for the entrenchment of the caste system is that it offers tremendous advantages to those in privileged positions, and so they have incentives to perpetuate it. However, this perspective misses an impediment to overcoming the caste system. Caste is so ingrained in the identities, culture, and religion of many Indian citizens that even many of those who are underprivileged, including those with ambitions for bettering their lives and those of their children, view marrying and many interactions with others outside their caste as abhorrent. Contact across some caste boundaries would be thought of as disgraceful to both sides.

## Homophily

*"Equals, the proverb goes, delight in equals . . . and similarity begets friendship."*

—SOCRATES, IN PLATO, *PHAEDRUS*

Such strong and sharp divisions among groups are not unique to India and its caste system, but appear throughout the world and for a variety of reasons.

The general tendency of people to interact with others who are similar to themselves was named "homophily" by Paul Lazarsfeld and Robert Merton (1954). The etymology of the word is clear— "homo" refers to same, and "phily" to love or fondness.

Figure 5.3 provides another example of homophily, in a very different context from the caste system in India. It shows how the friendships within a high school in the U.S. are divided along racial

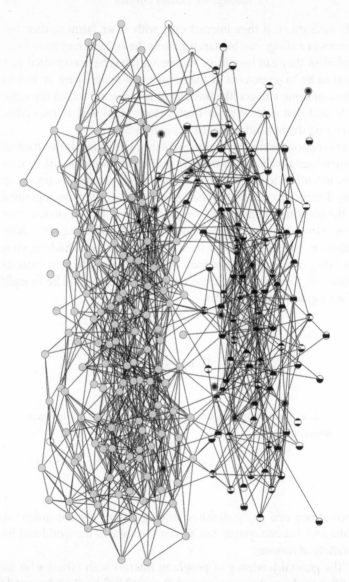

**Figure 5.3: Self-reported friendships in a high school from the Add Health data set. Students (nodes) with bold stripes are self-reported as being "Black," the nodes with center dots are "Hispanic," the nodes with light gray fill are "White," and the few remaining nodes are "Hispanic" (center dots) or "Other/Unknown (blank)." The figure was drawn using a spring algorithm, moving friends closer together and nonfriend pairs apart. The algorithm did not know the nodes' races, so the separation is the result of the friendship patterns.**

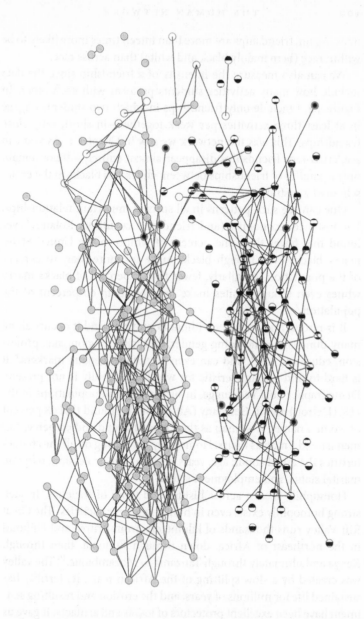

**Figure 5.4: The same network as in Figure 5.3, except now only "strong friendships," in which the two people did at least three things together in a week (hung out together after school, spent time together on a weekend, talked on the phone [data from the mid-1990s], etc.), are included. The split in the network is even more pronounced.**

lines. Again, friendships are more than fifteen times more likely to be within race (here mainly black and white) than across race.[4]

We can also measure the intensity of a friendship since the data include how many activities students perform with each other. In Figure 5.4, I include only friendships in which two students engage in at least three activities per week together—in short, only close friendships. This was the network we saw in Chapter 1. As you can see, cross-race friendships disappear almost entirely—there remain only a handful of friendships between whites and blacks in the entire school of 255 students.[5]

One can see such patterns in all sorts of important relationships. For instance, economist (and friend and coauthor) Roland Fryer found that fewer than one percent of whites in the United States marry blacks even though blacks make up more than 10 percent of the population.[6] Similarly, fewer than 5 percent of blacks marry whites even though whites make up more than 60 percent of the population.[7]

It is rare to find societies without homophily, and it occurs along many dimensions including gender, ethnicity, religion, age, profession, education level, and can even be seen in genetic markers.[8] It is hard to find a characteristic for which homophily is not present. Demographer Lois Verbrugge, in surveying adult populations in the U.S. (Detroit) and in Germany (Alt Neustadt), found that 68 percent of women named a woman as their closest friend and 90 percent of men named a man as theirs. Homophily was strong on all the characteristics she asked about: age, years of schooling, profession, religion, marital status, and employment status.[9]

Homophily is seen across history and types of societies. In fact, strong homophily exists even in hunter-gatherer societies. The Great Rift Valley runs thousands of kilometers, from Eritrea and Djibouti in the northeast of Africa, down through Ethiopia, then through Kenya and ultimately through Tanzania to Mozambique.[10] The valley was created by a slow splitting of the African plate. Its fertility has sustained life for millions of years, and the erosion and resulting sediment have been excellent protectors of fossils and artifacts. It gave us Lucy—a skeleton of a hominid ancestor of humans from more than three million years ago. It also preserved the earliest known tools, and animal bones with tool marks on them, also dating from between two and a half to three and a half million years ago. Remarkably, perhaps, there are still nomadic hunter-gatherer peoples inhabiting the Great Rift Valley, providing a lens into ways of living that date back millennia. One such society is the Hadza, who live around Lake Eyasi at the south end of the Great Rift Valley in Tanzania, neighboring the Serengeti. The roughly one thousand Hadza have retained their culture despite attempts by missionaries to convert them and encroachment by farmers and herders.

Theirs is not an easy life. Day to day, they move in accordance with available sources of food. They live mostly on fruits, honey, tubers, greens, occasional eggs, and sporadic meat, which they hunt with poisoned arrows. Obtaining calories is no easy task. Honey is often found by following birds that subsist on beehives often found dangerously high in tall trees. Hunting usually takes place in dry times, when animals are forced to gather around limited water sources. The lake is also subject to wild seasonal and annual fluctuations, sometimes full of water and hippopotami, and other times simply a dry and cracked bed of mud.

The unpredictability of available food also means that sharing is essential to staying alive. Coren Apicella and colleagues[11] studied the networks among the Hadza: who likes to be with whom in the next camp (as they break into fluid groups that merge and split depending on conditions), and who shares food with whom. Despite the free-form structure of the society, there was still significant homophily on many dimensions, including age, height, weight, body fat, and strength, even after controlling for other traits. For instance, an increase of 7.5 kilograms in how similar people are in body weight triples the probability that they are connected to each other.

Again, homophily is not surprising: there are many pressures that push and pull similar people together, in this case on physical and age dimensions. But this indicates that homophily appears regardless of the society in question. Indeed, at the other end of the technological spectrum, even when people can choose from an enormous pool of people, they still exhibit strong homophily, and the trend is even increasing.[12]

The Internet has led to an explosion of online dating and matchmaking. Fifteen percent of Americans report that they have used online dating sites,[13] and attitudes about the use of dating services have changed dramatically over time. The shift in attitudes toward and use of dating services by the young are especially striking, as more than a quarter of them have used an online dating site or app.

Such online sites provide for interesting dynamics in homophily. On the one hand people have many more choices and can reach out to people whom they would never meet in their day-to-day lives. These opportunities work to lower homophily. On the other hand, the technology allows people to filter and search, which allows them to be increasingly choosy as to which characteristics their potential dates and mates should have.

The filtering seems to win out, as we can see from more than 100,000 people's behavior on a German online dating site.[14] People on the site fill out a detailed questionnaire and then can also add their own profile, with pictures and text. People can then browse through the listings and send a message to anyone they would like to meet. Upon receiving a message, a person can decide whether to answer it. Men, on average, browse 138 profiles, send 12 first contact messages, and get 4 replies. Women browse 73 profiles, send 6 messages, and receive 4 replies. So, men have a third of their contacts reciprocated, while women have two thirds reciprocated.

We can see homophily in action by examining to whom people send contact requests, and which ones they answer, based on demographics. For example, when women send first contact messages out, they are 35 percent more likely than average to send it to a man who is of a similar educational standing, and 41 percent less likely than average to send it to a man who has a lower educational standing. Men are less picky—they are 15 percent more likely than average to seek out a woman of a similar educational standing, and 6 percent less likely than average to seek out a woman of lower educational standing. There are similar biases in the reciprocation rates, and the biases are statistically significant after controlling for other personal attributes such as age, height, and appearance.[15]

Looking at almost a million online daters in the U.S. shows that homophily is even stronger based on race, with both gay and straight users sending most of their messages to their own race, even when controlling for many other factors including education.[16]

When one group is much bigger than another, homophily's effects can be further amplified. As Chris Rock once joked, "All my black friends have a bunch of white friends. And all my white friends have one black friend."

Actually, even though this was a joke, it really is just simple accounting. Consider a group of nine whites and one black, and suppose that they are all friends with each other. Then each white has one black friend and the black has nine white friends. That's not too far from the relative ratios of the populations in the U.S., and so simple accounting will tell you that blacks are going to end up with more white friends on average than whites with blacks. That's true of any minority and majority group. Now add homophily to this necessary asymmetry and it can become very rare for a member of the majority group to have friendships with the minority group.

Guess how many black friends the typical white person in the U.S. has—where a friend is someone with whom that person "regularly discussed important matters"? *Zero*. Yes, at least according to one survey of over two thousand adults, three quarters of whites have *nobody* of any other race who is considered to be a close friend.[17] With such divisions, we can begin to understand why one group can be almost completely ignorant of another group's experiences, beliefs, and culture.

Homophily should not surprise you because, even if you have never heard of the word or any statistics about it, you would have had to have lived your life in isolation to have avoided it. However, the extent of homophily and its pervasiveness are striking and its stark divisions have many strong consequences for our exploration of the effects of human networks on our behaviors and outcomes.

## Location, Location, Location

*"We cannot marry Eskimos if there are none around."*

—PETER BLAU,
"CONTRASTING THEORETICAL PERSPECTIVES"

We connect disproportionately with others who resemble us for a variety of reasons, some involve our choices and others are due to things beyond our personal control.

Who better to offer advice than one who has experienced the same situation? New parents benefit from talking to other new parents about the plethora of choices they must make in raising their children. People studying to pass an exam to become a doctor or a lawyer or an actuary benefit from talking to other people who are studying for the same exam or have recently taken the exam. People in the same profession learn from others within their profession about new advances, techniques, and opportunities. Children gravitate toward others of the same age who have a similar level of maturity and common interests and concerns.

Beyond common interests, proximity has a huge influence on our friendships. Children in school spend most of their day with others who are within a year in age. Even if they wanted friendships across ages, their opportunities to have them are severely limited, as would be the benefits of being friends with someone they rarely see.

Natural divisions within businesses and other institutions, by expertise and tasks, also limit the contact that people have with others who have different training. Floor layouts in businesses are important in determining who speaks with whom on a regular basis. They are taken into account by architects who are planning anything from the design of courtyards in housing complexes to the workspace of a firm.

People also prefer to live in neighborhoods in which they find familiar businesses and culture. When people immigrate they actively choose areas in which others speak their language, celebrate their holidays, and practice their religion. Most immigrants are also heavily dependent upon friends and family in their new country to help them find places to live and work, and to navigate their new surroundings. Those connections are often to earlier waves of immigrants from a person's hometown. When Europeans migrated to the U.S., they even chose places with climates similar to those of their origin, as well as places where their skill sets fit with local industries. Wheat farmers made their way to the Great Plains, and meatpackers made their way to the stockyards of Chicago and Omaha. Swedes moved to Illinois, Iowa, Wisconsin, Minnesota; Norwegians made their way to the Dakotas and Montana; Germans to Illinois, New York, Wisconsin, Pennsylvania; Poles went to Chicago; and so on. For many years, Chicago had the second largest Polish population in the world, trailing only Warsaw.

Social networks are such an important initial lifeline that they end up heavily reflecting immigrants' heritage even a century later. The tightly knit communities based on ethnicity that formed during the heavy immigration periods in the late nineteenth century and early twentieth century still have an impact today. The density of a U.S. county's Irish or Italian ancestry is still a surprisingly strong predictor of the number of friendship ties on Facebook between that county and Ireland or Italy today. A location that has an extra one percent of people with ancestry from some country has a third of a percent more Facebook friendships back to that country today.[18]

Beyond co-location of immigrants by place of origin, people tend to live near others who are similar to themselves on other dimensions. For example, the co-location of a highly educated population is a not-so-secret secret of the success of Silicon Valley. Thirteen percent of the people living in Palo Alto hold a PhD, and that excludes Stanford's campus (where many faculty live and which is actually not part of Palo Alto). Nearby towns of Mountain View and Cupertino also have large numbers of highly trained and skilled workers. To find towns with similar fractions of highly educated people one usually has to look for college towns—either rural ones that capture the entire faculty and many graduates like Davis, California, or suburbs like Brookline, Massachusetts, which sits close to MIT and Harvard. Part of the success of Silicon Valley comes from the flow of ideas that one cannot avoid overhearing in any local coffee shop. Expertise and experience also move with a labor force that wanders freely from one company to another. The flow of ideas, information, and innovation, as well as the culture, in Silicon Valley makes it an obvious place for anyone starting a tech business to locate.[19]

Beyond information flows, there is an even stronger symbiosis that drives both the high-tech companies and highly skilled workers to locate in the same place. Chris Zaharias, the founder of a start-up named SearchQuant, worked for Netscape, Efficient Frontier, Omniture, Yahoo, and Triggit, before starting his own company. That sort of résumé is not unusual. Modern companies appear and disappear quickly, especially high-tech ones. If these companies were spread all over the world, then employees of these companies would have to move every few years. That would be costly for both the employees and ultimately the firms. By living in Silicon Valley you can work for a company that is about to disappear and can count on landing in a new company within a few miles of home even before the old one closes its doors. As both tech workers and companies flock to Silicon Valley, the concentration of people with similar backgrounds becomes an even stronger attractor and it becomes difficult for anyone serious about a high-tech career or business to locate anywhere else.

This is not unique to Silicon Valley: the same effect drives Hollywood's long-standing grip on the movie business, as well as the high concentration of the financial industry in cities like New York, London, Tokyo, Singapore, and Shanghai. Thus, tendencies for co-location of people who work together and have similar skills is another force behind homophily. There are other aspects of geographic location that can amplify homophily.

## Schelling's Insight

Thomas Schelling won the 2005 Nobel Prize in economics for his fundamental contributions to game theoretic analyses of conflict and coordination. One of his many profound insights was to discover a strikingly simple and yet subtle force behind segregation.

The fact that cities are segregated by ethnicity and income is evident even to the casual observer. This is a form of homophily and many of the forces that shape it include the ones we discussed above.

What Schelling pointed out, however, was that even *small* biases in preferences over the ethnicity of one's neighbors can have an enormous impact. Schelling's model is remarkably robust. It has driven many follow-up variations and studies.[20]

Schelling's model works as follows. People come in varieties, say according to ethnicity or religion or caste, or some combination. To keep things simple, let's look at an example in which people are either "solids" or "checkered." I will refer to them as families, who live on a grid. Each family has up to eight neighbors occupying the squares immediately adjacent to them (on their sides, above or below them, or on the diagonals adjacent to them). Each family has some preferences over the ethnic composition of their neighbors. A family is happy as long as at least one third of their neighbors are of the same type as they are.

This is a very weak restriction on preferences: people simply prefer not to be too small a minority in their neighborhood. If people had stronger preferences, with everyone wanting to be surrounded by others who are of the same type, then it would not be surprising to get separation. Instead, in Schelling's model, people only require that some minimal fraction of their neighbors be similar to them in order to be happy.

Let's look at an example of what happens, starting with thirteen families randomly placed on a grid as pictured in Figure 5.5. The blank squares are unoccupied. Families are happy to stay in their current location as long as at least one third of their neighbors are of their type, but otherwise they wish to move.

**Figure 5.5: Schelling's model of segregation. We begin with an integrated city consisting of solid and checkered families, and a few unoccupied (blank) squares. A family is happy as long as they are not too small a minority in their immediate neighborhood. If fewer than one third of their neighbors are of the same type, they would like to move.**

There are three unhappy families in Figure 5.5, each is marked with an X in Figure 5.6. For example, the checkered family near the bottom has three fourths of its neighbors solid and only one fourth checkered.

In Schelling's model, an unhappy family moves to an unoccupied (blank) square at random. Let us do this one by one. There is nothing very strategic about it, families do not anticipate the future of the process, they just move to some open spot if they are unhappy. If you want, you can enrich the model so that unhappy families look to find a new location in which they would be happy, but the model works even without that enrichment. People keep moving around until all of them are happy. Once they are all happy, they stop and we can see what the resulting pattern looks like.

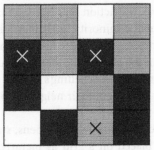

**Figure 5.6: The three families with Xs are unhappy: each has fewer than one third of their neighbors of the same type. The other families start out happy.**

Let's give it a try. Figure 5.7 shows us a series of moves. In panel (a) of Figure 5.7 one of the unhappy families moves. In this case it is a solid family moving down and to the left, leaving its old location (where the x is) empty. The choice of which family moves first is random.

Note that as we move the solid family as in panel (a) of Figure 5.7,

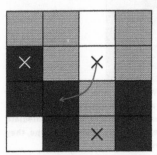

(a) An unhappy solid family moves down and to the left.

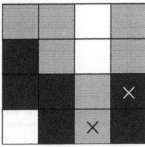

(b) Now there are two unhappy families.

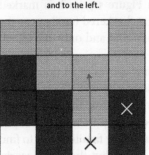

c) An unhappy checkered family moves.

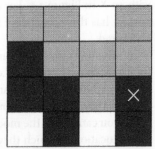

(d) There is still one unhappy family.

**Figure 5.7: Unhappy families move. The order is random, as are the spaces that they move to.**

it changes which *other* families are unhappy. As families move they increase the presence of their type in their new neighborhood and decrease the presence of their type in the old neighborhood. That affects which others are happy and unhappy. You can see this effect by comparing the unhappy families from Figure 5.6 to panel (b) of Figure 5.7. The unhappy solid family at the very left of Figure 5.6 did not move but it gained a solid neighbor and so is now happy. A different solid family at the right of Figure 5.6 was originally happy, now becomes unhappy in Figure 5.7 panel (b) since it lost one of its solid neighbors.

These externalities are an important component of the Schelling model. As families of a given type leave a neighborhood, that prompts others of their type who are left behind to become unhappy and move as well. It can also make people in the neighborhood that they move to who are of other types unhappy. This cascades quite dramatically.

After the moving is over, we are left with the configuration in panel (b) of Figure 5.8. We end up with strong separations between solids and checkers even though all families were happy to live in integrated neighborhoods; they just wanted to be sure that some minimal fraction of their neighbors was like them. The model works for many different levels of preference and for various forms of geographies.

The forces underlying Schelling's model should be becoming familiar: it involves externalities and cascading behavior. People's decisions to move affect their neighbors' happiness. The feedback effect leads to cascading moves, and small initial biases in preferences get amplified via the interactions and have dramatic consequences.[21]

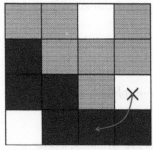

(a) The remaining unhappy solid family moves.

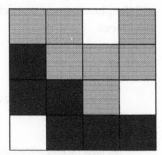

(b) Everyone is happy.

**Figure 5.8: Even though every family was willing to be a minority in their neighborhood, just not *too small* a minority, the resulting moves led to a very segregated community.**

One study estimated the thresholds for "white flight" by examining how large a percentage of minority populations needs to move in in order for many whites to leave. They found evidence of thresholds of between 5 and 20 percent—so that whites exited even when small numbers of minorities moved in.[22] This suggests that those moving out actually had preferences to be in a large majority, which would be a much more extreme situation than our Schelling example. This might also reflect effects similar to those that we saw in the Keynesian Beauty Contest: people might move in anticipation of others leaving the neighborhood, adding another layer to Schelling's insight.

Of course, a simple model like Schelling's is bound to miss some aspects of reality. Some neighborhoods are highly racially mixed and stay mixed, while other neighborhoods are more reactive and shift with even small increases of minority populations.[23] Not everyone has the same preferences: some prefer to live in very homogeneous neighborhoods and small changes in their neighborhoods lead to dramatic moves, while others are happy to live, or even prefer to live, in very heterogeneous neighborhoods.

Still, Schelling's model provides important lessons. It makes clear that even relatively minor preference biases can have a large impact given the cascades that can result from the externalities involved in people's decisions of how to situate themselves—and you can see that this applies not only to neighborhoods, but also to clubs and networks. This helps explain the strength of homophily that we see across many dimensions. There are many forces that push to homophily, and the phenomenon that Schelling points out can amplify any and all of them.

## More Reasons for Homophily

*"The most exciting attractions are between two opposites that never meet."*

—ANDY WARHOL

You might expect that the forces we have already discussed would cover most of the drivers of homophily, but it is so pervasive because of the multitude of forces conspiring to push us toward our brethren.[24]

Community boundaries are important in allowing us to cooperate and trust each other in our everyday lives. Repeated interactions allow us to reciprocate when people help us out and to punish detrimental behaviors. We can trust our neighbors to look out for our children and other prized possessions, as we can do the same for them. We even teach our children to beware of strangers.

As human populations become increasingly urbanized, defining those communities becomes increasingly challenging. We build networks around our professions, ethnicities, religions, and other common features that put us into more frequent and well-defined contact with each other. Instead of local geography defining our communities, we rely more directly on homophily and repeated contact to define our trusted circles of friends.

Homophily also reinforces itself. People can better predict the behaviors and reactions of those close to them.[25] They better understand the local culture and norms, and how they are expected to act in various circumstances. Although it lowers stress and helps people coordinate in their day-to-day lives,[26] it also ends up increasing differences across groups and making it relatively easier and safer to interact in one's own spheres.

Our understanding of homophily would also be incomplete if we failed to account for competition between groups. People reaching across groups can be seen as traitors or with suspicion. Who can forget Romeo and Juliet, and the many other feuds and gang wars that have plagued societies throughout history. Competing groups often exhibit strong rivalries, from the clan divisions in ancient China to the schism between Sunni and Shia in the Islamic world. It is hard to find a part of the globe with substantial ethnic or religious heterogeneity that does not have rivalries.

As animosity and suspicion between groups grows, so does homophily. Prior to 2004, segregation in Spain between native Spaniards and Arab immigrants had been declining. The morning of March 11, 2004, abruptly reversed that trend. Ten bombs exploded on four different trains in the middle of rush hour, killing hundreds and wounding thousands of people. The investigation was complicated and followed many false leads, but eventually led to the conviction of more than twenty people. The Islamic militancy of the core cell of terrorists behind the bombing led to heightened tensions between natives and Arab immigrants. It has been estimated that segregation between the Arab and Spanish population increased by more than 5 percent in the following two years—a large amount considering how infrequently people move.[27] This is certainly not unique to Spain, and the impact of any single event can eventually dissipate. Still, it shows how important and salient perceptions of group identities can be.

Understanding how easily people assume roles and identities provides further insight into chasms between groups based on identities, another driver of homophily. Even though by now it has become almost a cliché, the famous Stanford Prison Experiment provides a frightening example of how simply having people assume identities can lead to divisions and even conflict between groups. A Web site dedicated to its description begins with the following introduction:[28] "On a quiet Sunday morning in August, a Palo Alto, California, police car swept through the town picking up college students as part of a mass arrest for violation of Penal Codes 211, Armed Robbery, and Burglary, a 459 PC. The suspect was picked up at his home, charged, warned of his legal rights, spread-eagled against the police car, searched, and handcuffed often as surprised and curious neighbors looked on. The suspect was then put in the rear of the police car and carried off to the police station, the sirens wailing."

This experiment was headed by Philip Zimbardo, Craig Haney, and Curtis Banks in 1973. The subjects of the experiment were twenty-one healthy college-aged males who were arbitrarily split into two groups: guards and prisoners. The prisoners had not really committed armed robbery or burglary and the guards had no prior roles in law enforcement. A makeshift prison was built in the basement of a Stanford building with the advice of former prison inmates. After being "arrested" in this dramatic fashion to help them assume the role of prisoners, the subjects who were picked to be prisoners were searched, stripped, and sprayed to delouse. They were given prison clothing and assigned to small cells without clocks or windows. The guards were not given any particular guidelines, instead, as described on the Web site, "they were free, within limits, to do whatever they thought was necessary to maintain law and order in the prison and to command the respect of the prisoners."[29]

What ensued was far beyond what the designers of the experiment had anticipated. The subjects became so ingrained in their identities and roles that the experiment took a dangerous turn. By the second day the prisoners began to rebel, yelling at the guards about their treatment. The guards responded by spraying the prisoners with fire extinguishers, removing beds from the cells, and putting the leaders of the rebellion in solitary confinement. On that second day of the experiment, one of the prisoners, designated #8612, broke down with hysterical ranting and crying. The guards thought he was acting and refused to release him, instead offering him improved treatment if he would act as an informant. When his situation worsened, it became clear that he was not acting and the subject was released from the experiment. By the fifth day of the experiment it had to be ended. Some guards were seen to be "acting sadistically" and others were not able to control them, and the prisoners were seen to be "withdrawing and acting in pathological ways."

## Some of Homophily's Implications

Alberto Alesina and Ekaterina Zhuravskaya[30] examined how segregated countries are along ethnic, linguistic, and religious dimensions and how this relates to the success of the country according to various measures. They work with a "segregation index" that is a variation on a homophily index—comparing how concentrated populations are within various regions in a country compared to what things would look like if people were distributed across regions in proportion to their presence in the population. So, for instance, in South Africa, roughly 80 percent of the population are black African, 9 percent are locally referred to as coloured (a variety of mixed ethnic origins), 9 percent are white, and the remaining are of Asian origin. If there was no segregation, then each region would have roughly this same mix of population. On the other hand, for example, if some regions are almost all black, and other regions are all coloured, and others all white and Asian, then there is substantial segregation. So, a score of 0 on the segregation index means that different ethnic groups are smoothed evenly across all regions. A score of 1 means that different ethnicities live in entirely different regions.

The five most segregated countries are Zimbabwe, Guatemala, Afghanistan, Uganda, and Turkey, all scoring .36 or above. The least ethnically segregated countries are Germany, Sweden, the Netherlands, Cambodia, and South Korea, all scoring less than .01. China, India, Russia, Israel, and Spain are in the .08 to .24 range—and the U.S. and U.K. are in the .01 to .03 range. That might surprise you, but these are fairly large regions and the segregation within many U.S. and U.K. cities is at a neighborhood level. This measure of segregation is capturing whether different ethnicities live in entirely different regions of a country rather than whether there is homophily at a local and network level. For example, Arabs and Jews tend to live in different cities in Israel, with some exceptions, and so it has a high segregation score. In contrast, Asians, blacks, whites, and Hispanics often live in many of the same cities in the U.S., but in different neighborhoods. So Israel shows up as being much more segregated than the U.S.

Regardless of why a country might have a low or high level of segregation, we can still see whether this crude measure of segregation predicts how well a country functions. Just from the names of the countries and the levels of segregation, you can guess where this is going. The most segregated countries are the poorest and have the worst-functioning governments; and the least segregated countries are the wealthiest and have the best-functioning governments.

Using the Alesina and Zhuravskaya data, I have plotted out how a few basic variables correlate with a country's segregation index. To keep the figures uncluttered, I have only included countries with a minimal GDP. Ghana just makes my cut of minimal GDP and Côte d'Ivoire falls just short. The figure with all countries actually has an even steeper relationship, but more labels become unreadable.[31]

In Figure 5.9 we see how a country's overall productivity, in terms of GDP per capita, relates to the country's segregation. The basic trend is that higher segregation leads to lower productivity. The average GDP of the least segregated countries is more than six times as high as the average GDP of the most segregated countries. The relationship represented by the best-fit line through the data is negative

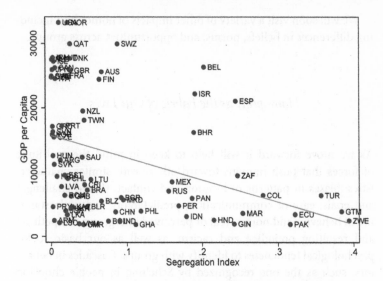

**Figure 5.9: Countries' GDP as a function of how segregated they are.**

and highly significant in a statistical sense (at more than the 99.9 percent level).

There are other things to note about the relationship in Figure 5.9. It is still very noisy: there is lots of variation between countries. For instance, countries that have very low segregation have a wide variety of GDP levels. Having low segregation is no guarantee of having high GDP: there are societies with very low segregation that do very poorly. What is important to note, however, is that at the highest levels of segregation—above .25—there are no countries with high GDP. Having very high levels of segregation seems incompatible with high GDP: there are no countries in the upper right part of the figure (that are highly segregated and have a productive economy). Low segregation does not guarantee prosperity, but high segregation seems to preclude it.

Why might greater segregation lead to less functional governments and economies?[32] Alesina and Zhuravskaya offer an explanation: greater segregation provokes lower trust between people in the country, and makes for the proliferation of special-interest political parties, which can then lead to more divisive politics and less effective governments.

We will soon visit a variety of other impacts of homophily, including differences in beliefs, norms, and opportunities across groups.

## Homophily as the Fabric of Our Lives

As we move forward it will help to keep in mind the multitude of forces that push humans toward others with similar characteristics: biases in patterns of location and contact, shared challenges and goals, ease of communication, predictability and understanding of behavior and norms, real or perceived intergroup competition and resulting prejudice and racism, as well as our basic sociopsychological tendencies to identify with groups. Cascades in behaviors, such as the one recognized by Schelling in people choosing where to live, amplify the divides among groups; and feedback from diverging norms and behaviors reinforces and deepens our schisms.

Homophily is so fundamental to human networks and we are so familiar with it that it fades into the background. We lose sight of it not only in our day-to-day lives, but also when we look for explanations for our culture, norms, and behaviors. While homophily makes our lives and the behavior of those around us predictable,[33] it is also very divisive. The splits in our social structures are so deep, pervasive, and resilient that they play primary rather than secondary roles in understanding polarization in beliefs and opinions as well as persistent inequality in opportunity, employment, and well-being, as we shall see in the chapters that follow. Forgetting about these basic network divides can lead one to propose ineffective policies when one tries to fix such problems.

## 6 · IMMOBILITY AND INEQUALITY: NETWORK FEEDBACK AND POVERTY TRAPS

### *Immobility*

Claire Vaye Watkins grew up in Pahrump, Nevada. Pahrump means "water rock" and it sits midway between Las Vegas and Badwater (Death Valley). As goes with the region, the climate can be brutal and the average temperature is over 100 degrees in the middle of the summer. Pahrump is remote but has valuable water sources— springing from artesian wells.[1] It was home to the Southern Paiute before being settled by pioneers in the 1800s. Its cotton and alfalfa production have given way to other sources of income, and today it boasts several casinos, (legal) brothels, a golf course, and a speedway. The median income in Pahrump lies well below the U.S. median, but places it solidly in what one might call the lower middle class.

You might expect that the very highest-achieving students from a lower-middle-class community would have a chance at ending up in a top college. However, as Watkins described in an essay called "The Ivy League Was Another Planet," you would be completely mistaken.[2] She describes a trip to Lake Tahoe when she was in twelfth grade. She, and a friend from Pahrump named Ryan, had been selected as finalists for a competition to choose the "Top 100" seniors in the state of Nevada. As she describes it:

> On the plane, Ryan and I met a boy from Las Vegas. Looking to size up the competition, we asked what high school he went to. He said a name we didn't recognize and added, "It's a magnet school." Ryan asked what a magnet school was, and spent the remaining hour incredulously demanding a detailed account of the young man's educational history: his time abroad, his after-school robotics club, his tutors, his college prep courses.
>
> All educations, we realized then, are not created equal. For

Ryan and me, of Pahrump, Nev., just an hour from the city, the Vegas boy was a citizen of a planet we would never visit. What we didn't know was that there were other, more distant planets that we could not even see. And those planets couldn't see us, either.

Even though she lived only an hour from a major metropolitan area, Claire knew precious little about colleges and how to get into them. Her sense of isolation and local lack of knowledge help explain why immobility and inequality are not temporary phenomena. Claire's story begins to answer the question of why more people don't educate themselves in spite of the increasing wage polarization between the educated and the uneducated.

Immobility results when people end up trapped by the social circumstances into which they are born: the networks in which they are embedded fail to provide them with the information and opportunities that they need to succeed.

Beyond moral objections, immobility is problematic because it is inefficient: it lowers productivity as some of a society's most productive individuals end up locked in unproductive roles. How many Picassos have spent their lives working in coal mines? What if the person who would have otherwise found a cure for a cancer was just born in a slum? As such, immobility can have a substantial impact on a country's growth rate.

Homophily plays the key role in our story of immobility. Homophily shapes the information that parents have, and the ways in which they raise their children. Even controlling for parents, the community in which a child exists goes a long way toward explaining that child's educational and lifetime earnings opportunities. It affects what they expect from society and what a society expects of them. Alona King, a young black woman who majored in computer science at Stanford, captures the point: "I hate walking around the halls of the Gates Computer Science Building.... I always get hit with the four words every non-tech minority thinks whenever they see an unfamiliar minority in a tech space: 'Hey, are you lost?' "[3]

In order to develop an understanding of the full role of networks in immobility and inequality, we will begin with some background on inequality and immobility, and their tight interrelationship. We will also discuss the changing labor market—the growing distance between wages of people with and without a college degree. The final piece of background is immobility in education—children born to uneducated parents and networks tend to end up with lower levels of cognitive interaction and education from birth, and much lower chances of any higher education.

Once we have this background in place, networks will then enter the story in several ways. First, as we see in Claire's story above, homophily limits information flows and basic knowledge not only of the value of education, but how to go about getting it. This is true of both parents choosing how to raise their children and what the children learn from those around them. Second, incentives to get an education and to enter the formal labor force are heavily dependent upon the behavior of one's peers and community. Third, wages and opportunities once one is in the labor force are heavily dependent upon how many of one's peers are employed since networks of contacts are the primary source of getting a job. These network effects when combined, especially given that they feed off of each other, help us to understand persistent and increasing immobility and inequality. These forces can be more powerful than wealth effects from parents to children, which explain only a fraction of recent trends.

## Immobility, Inequality, and the "Great Gatsby Curve"

*"So we beat on, boats against the current, borne back ceaselessly into the past."*

—F. SCOTT FITZGERALD, *THE GREAT GATSBY*

It is clear that a child's success and happiness later in life depend upon the start afforded to her by her parents and community. Standard ways of measuring the stickiness of status across generations is to compare children and their parents according to factors such as income, wealth, and college achievement. How strongly correlated are the fates of children with their parents'? Figure 6.1 shows how

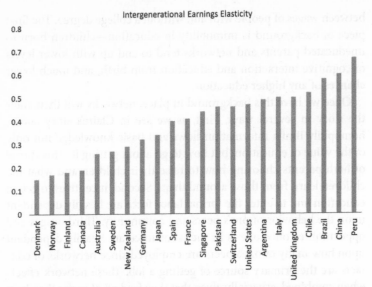

**Figure 6.1: Immobility: Intergenerational earnings elasticity. This is a normalized measure of the correlation between child and parent earnings, with 0 meaning that child and parental earnings are essentially uncoupled, and 1 indicating that a parent's relative standing in earnings is fully reflected in the child's relative standing. Data from Miles Corak (2016).**

a child's eventual income relates to that of her parents. The specific measure pictured here is what is called "intergenerational earnings elasticity."

Intergenerational earnings elasticity can be thought of as answering the following question. Suppose that your family earns 10 percent more per year than my family. How much more do you end up earning than I when we grow up? If the answer is 10 percent, then the entire advantage that your family had over mine is preserved across generations. Since the ratio of how much more you earn than I compared to how much more your family earned than mine is 1, there is complete immobility: we end up with the same relative rankings as our parents. If instead you earned just one percent more than I do, then only one tenth of your family's advantage over mine would have been passed along. If we have the same incomes, then there is complete mobility as our relative standings are independent of our parents' relative standings, which corresponds to an elasticity of zero.

Thus, the elasticity is a measure of immobility. The U.S.'s elasticity/immobility is just under one half, and Peru tops the chart at two thirds.[4][5]

You might find the United States' position in this graph surprising. For instance, it has more than double the immobility of Canada. The U.K. has a similar position.

In contrast to how surprising the high immobility in the U.S. is, high *inequality* in the U.S. is self-evident. You only have to cross any major city and you cannot avoid passing through both very poor and very rich neighborhoods. But the "American Dream" is ingrained in our psyche and identity, as is the idea that the U.S. is the "land of opportunity." People are rewarded for their hard work, skills, and talents, not their bloodline. As the unalienable right of the pursuit of happiness is central to our Declaration of Independence, one tends to expect that the U.S. would at least compare well with other countries on a (im)mobility scale.

We all know of incredible success stories. My father managed to put himself through college and graduate school, and to become a nuclear physicist, even though his father had not completed high school and worked for a trucking company. Historically, the land of opportunity was an appropriate moniker. During immigration waves of the nineteenth and early twentieth centuries, the country was expanding westward and experiencing extended economic booms (with some hiccups). Opportunities for employment and growth were abundant, especially for immigrants, many of whom were escaping poverty in the countries they were fleeing. People were moving to California and striking it rich. Education was valuable, but a less essential key to unlocking what was then a decent living. It was a golden age for growth, and the American Dream. In another period, after the Second World War, when manufacturing and the economy boomed, and the middle class flourished, climbing the economic ladder was also quite feasible.

However, by the time I started studying immobility about fifteen years ago, things were strikingly different. Upon first seeing parent-child correlations in education in the U.S., I thought that there must have been something wrong with the data. After some digging, I understood that all the figures painted the same story—the U.S. now ranks very high on the scale of parent-child correlations regardless

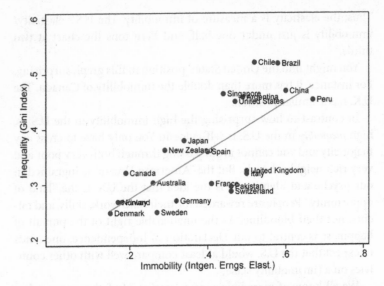

**Figure 6.2: The Great Gatsby Curve: a strong relationship between inequality (as measured by the Gini index of each country on the vertical axis, with 0 indicating no inequality and 1 indicating maximal possibly inequality (more on this below) and immobility (as measured by a country's intergenerational earnings elasticity) across countries. Immobility data are from Corak (2016); the Gini indices are from the *CIA World Factbook*.**

of whether one measures income, education, wealth, or even longevity. Ranking high on these scales is not a good thing—it means high *immobility*.

If you expected the U.S. to rank well on economic mobility, you are not alone. Many people's perceptions of mobility and reality are mismatched. For instance, in Figure 6.1, Germany comes in at .32 while the U.S. has a rate of .47, almost one and a half times the rate of Germany. Yet, if you ask people whether they agree with the statement "If I work hard, I can improve my life," then you find that almost 84 percent of Americans agree while only 62 percent of Germans agree.[6]

The relationship between inequality and immobility was crystallized by Alan Krueger, who gave a famous presentation in January of 2012 when he was the chairman of the Council of Economic Advisors. Krueger presented the Great Gatsby Curve: the strikingly strong relationship between inequality and immobility that had been noted by Miles Corak in 2011. Krueger, in trying to find a name for the curve, offered a bottle of wine to the junior staffer who came up with the best name—a prize won by Judd Cramer, who named it after Gatsby's difficulty in jumping social classes.[7] The tight relationship between inequality and immobility is pictured in Figure 6.2.

It is important to note that some aspects of the relationship between immobility and inequality are mechanical. If we compare a small and homogeneous country to a large and heterogeneous one, then we expect to find lower inequality and immobility in the small homogeneous country, all else being equal. For instance, if we only examined the children of engineers, we would see small amounts of inequality as most children would end up highly educated and well-paid, and also low correlation between parent and child income as the small differences in parents' income would be mostly inconsequential in determining which children fared better or worse. The same would be true if we only examined the children of coal miners—there would be low inequality, as most would end up in the working class, and there would be almost no correlation in outcomes between parent and child since the parents would have almost identical characteristics. Much of immobility and inequality comes from the comparison across different segments of an economy, and so as an economy becomes more diverse then we should expect to see both more immobility and more inequality.

Nonetheless, the Great Gatsby Curve is picking up much more than a mechanical size effect. For example, its small size explains part of why Denmark is low on both immobility and inequality, but not all. Low-skilled workers in Denmark earn about 50 percent more than people in comparable jobs in the U.S., and they enjoy many more benefits. Thus, the lower inequality in Denmark compared to the U.S. derives in part from the fact that the relatively poor unskilled workers in Denmark are much better off than people with similar skills in the U.S. In a society in which most people can choose whether to work in lower- or higher-skilled jobs, there is a natural force that pushes wages of lower-skilled jobs up: one then has to entice workers to take such a job rather than having a large captive unskilled labor force with no other opportunities. In terms of immobility, one sees that the correlation in educational attainment between parents and children in Denmark is less than 10 percent, while in the U.S. it is almost 50 percent, despite the fact that they have similar percentages of the population attending college—so it is not just homogeneity driving the different positions on the curve.[8]

Almost a quarter of Canada's population consists of various minorities, and the country has sizable immigration, and a very diverse economy in terms of the spectrum of goods and services it produces. Yet, still it is among the most mobile countries in the world, and has inequality that is not too far behind that of the Nordic countries. What is interesting about the Great Gatsby Curve is why a country like Canada is low on both scales, while the U.S. and China are so much higher on both scales.

To emphasize that inequality is implied by immobility, I have flipped Krueger's original Great Gatsby Curve. In Figure 6.2, immobility is on the horizontal axis and inequality is on the vertical axis—so we are viewing inequality as a function of immobility. The logic is that a substantial portion of inequality results from immobility. If all children were born into similar networks with the same opportunities of advancement—with similar parents, peers, and communities—then the only inequality would result from some random variation in their innate talents and personalities, as well as their choices, and some luck. If instead, children are born into very different networks—with different parents, norms, information, and opportunities—the resulting inequality is more extreme. So, part of the inequality in outcomes is driven by the difference in opportunities and mobility.

Ultimately, immobility and inequality are deeply intertwined: inequality also begets immobility, and there is feedback. The differences in parents, peers, and communities that constrain opportunities are due in part to inequality. But ultimately the entrenched networks of information and norms that are shaped by homophily constrain opportunities and behaviors, and so it is these forces behind immobility that are primary—and one can view inequality as a result and not as the root cause. This is different from the usual story that wealthy parents can afford to do more for their children.

This perspective is especially helpful in thinking about useful ways of breaking a vicious cycle and improving both immobility and inequality. It shifts the spotlight away from things like the tax rate on capital, instead to thinking about how to deal with the foundational social structures that drive the resulting immobility and inequality.

## *Inequality*

*"In a country well governed, poverty is something to be ashamed of.*
*In a country badly governed, wealth is something to be ashamed of."*

—CONFUCIUS

Reducing inequality to a single score or index inevitably over-simplifies, just as we saw with measures of centrality and influence. Should we be measuring wealth, income, expenditures, consumption, employment, happiness, longevity, or some combination of these? How should we account for taxes and welfare payments? Should we look at families or individuals? How do we account for numbers of children? There are so many things to measure, but many of these are highly correlated, so the good news is that looking at some simple indices paints a pretty robust big picture.

Corrado Gini, an Italian statistician, demographer, and sociologist, developed a succinct and yet insightful measure of inequality in 1912. It is now known as the Gini index. It is ironic that Gini was a fascist and eugenicist. Gini was an early supporter and confidant of Mussolini. He did eventually have a falling-out with the fascist movement, but mainly because it meddled with his academic pursuits. Nonetheless, he developed what has become the most widely used measure of inequality

An easy way to think of the Gini index is as follows.[9] Pick two people from the society. How much more income does the richer person have compared to the poorer of the pair? If we see how big this is on average across all possible pairs then we are measuring how unequal the society is, on average. To be able to compare across countries we'll normalize by twice the country's average income. The "twice" makes sure that the Gini lies between 0 and 1 since we are comparing pairs of people: The extreme unequal case is one in which we take all of one of the pair's income and give it to the other, in which case the difference of incomes between them is twice the average income; so dividing by twice the average normalizes this most extreme case to be 1. If everybody has the same income in a society the Gini is 0.[10]

Figure 6.3 presents Gini indices of a selection of countries. Ginis generally range from .25 in very equal societies to .7 in extremely unequal societies, depending on the time in history and exactly how one measures income.[11]

In these data, Denmark is the most equal society, with a Gini of about .25, and South Africa ranks as the most unequal among these countries with a Gini index of over .60, well over twice as unequal.

Just as we see inequality varying across countries, it has also varied over time. Clear pictures of income distributions going back centuries are hard to obtain, but with some careful detective work and extrapolations, people have put together estimates.[12] For example, although one might imagine that historic hunter-gatherer societies would be idyllic in terms of equality, they also exhibited significant inequality.

One historic hunter-gatherer society bordered Keatley Creek, which carries water down from the mountains of British Columbia. It lies a couple of hundred kilometers northeast of Vancouver. For

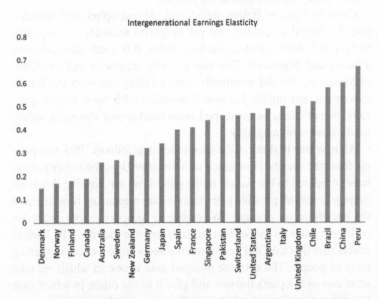

Figure 6.3: Gini index data from the *CIA World Factbook*. The Gini index, ranging from 0 (no inequality) to 1 (extreme inequality), shown for selected countries.

thousands of years Keatley Creek was home to aboriginal societies that lived off the salmon that populated the river from spring to fall, as well as deer, fox, bear, and other animals that inhabited the forests of the region, and a variety of natural tubers, wild onions, raspberries, and other local plants. Although wild game can be abundant in the area, having access to the best sites makes a difference. For instance, there are some sites along the creek that are much better for fishing than others. Some families controlled access to prime sites, and that control was maintained across generations. Excavation of various longhouses suggests that the families who controlled the prime fishing and hunting sites had substantial storage of a wider variety of foods and large areas for sleeping and large hearths for cooking and heating, while other families lived nearby with relatively little storage and cooking.[13]

This sort of inequality is not just anecdotal. As soon as humans began accumulating goods and exerting exclusive control over land, they began amassing wealth and transmitting it across generations. By examining various pastoral and horticultural societies, anthropologists have found a wide range of Gini indices in land and livestock holdings, from .3 to .7—with averages in the .4 to .5 range.[14]

Although inequality has been around for millennia, it really soared during the Industrial Revolution. The Industrial Revolution had unprecedented growth in productivity. Because of this growth, combined with poor regulation of monopolies, the top few in the industrializing nations gathered unmatched percentages of their countries' wealth. As economies grew, inequality levels began to reach all-time highs. By the middle of the Industrial Revolution, England and Wales were estimated to have a Gini coefficient of .59 and Europe in general around .57. These are not far from the extreme of modern South Africa. The United States started out somewhat more agrarian and egalitarian, with an estimated Gini of .44 in 1774 just before its birth. However, by the time of the Civil War its industry was also emerging and the U.S.'s inequality began catching up to Europe's, hitting a Gini of .51 by 1860 (including slaves). By the early twentieth century, the U.S. had overtaken much of Europe not only in industrial production, but also in terms of inequality—rivaled only by prewar Germany.

Countries that industrialized more recently, such as China, India, and parts of Africa, had lower Gini indices during the Industrial Rev-

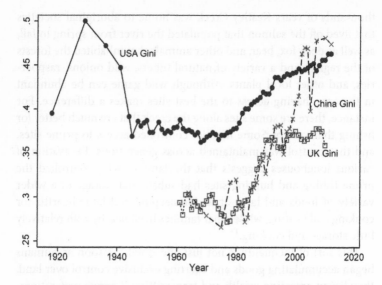

Figure 6.4: Gini indices for China, the United Kingdom, and the United States over time. Data for China and the U.K. are from the World Bank's "All the Ginis" data set created by Branko Milanovic, version November 2014. Data for the U.S. are from Atkinson and Morelli's 2014 "Chartbook of Economic Inequality."

olution. As they have industrialized, their Gini indices have soared as well, as the richest segment of the population has gathered much of the newfound income, while many others are still living near the poverty level. Overall incomes have been rising, but much faster at the top. The rapid ascent of China's Gini index over the past few decades is seen in Figure 6.4.[15]

Several factors temporarily counteracted the extreme levels of inequality that emerged with the Industrial Revolution. Unions grew, as did antitrust legislation and the regulation of monopolies by governments. Moreover, larger governments—spurred by the Great Depression and two world wars—increased taxation and redistribution. In addition, manufacturing technology required a larger labor force with new skills and the demand for labor pushed employment to new highs. A middle class emerged. In the decades following the Second World War, especially in Europe and North America, the middle class experienced unprecedented growth in its standard of living. The middle class not only gained in terms of the goods that were increasingly within their reach—from telephones to cars to radios and televisions—but the middle class also made huge gains as more of its members went on to higher education.[16] Gini indices dipped to some of their lowest points since the Industrial Revolution. We see that dip in Figure 6.4.

From roughly the 1980s to the present, several trends led to a resurgence of inequality in many countries around the world, including many not appearing in Figure 6.2. One trend is how technological change—instead of increasing demand for labor—is replacing and displacing labor. This is having a huge impact on the middle and lower middle classes in many countries. As technology has become increasingly sophisticated, it is able to replace labor in the provision of more and more goods and services. This does not mean that it puts people permanently out of work, but rather that now we have fewer automobile workers and travel agents and more dog walkers.

This transition can become ever more painful if not properly addressed. In the 1940s, Joseph Schumpeter coined the term "creative destruction" to describe the process by which new innovations make theretofore prevalent technologies and companies obsolete. We are witnessing it now on an unprecedented scale. Many innovations are replacing labor, except at the highest skill levels. Although obsolete machinery can be junked or recycled, what does a society do with its labor?

As an example, London taxi drivers have traditionally had to pass one of the most difficult tests on facts in the world, known as the Knowledge, in order to become licensed black-cab drivers. Many would spend years studying the complex map of London, memorizing tens of thousands of small streets, alleys, and landmarks, and learning to provide optimal routes between points quickly from memory. If you have ever taken a ride in a cab in London, it can be impressive. But such knowledge has been made obsolete by GPS technology and mapping apps—which not only calculate routes in an instant, but also adjust for current traffic conditions. This has led to contentious exchanges between the cab drivers and some of the British politicians who have suggested that the time for such exams is past.

Such advances in technology have lowered the labor needed in all types of production, from agriculture, to manufacturing, to services. Most travel bookings are now done online. Even famous retail stores are disappearing as online shopping takes over. We are on the verge of driverless cars and trucks, which will further transform the transportation industry. The recent effects of this have been most pronounced on the middle and lower middle classes—who have traditionally provided much of the labor in manufacturing and services that require a modicum of skills and experience but not higher education.

Figure 6.5 shows how much change in labor productivity there has been in the last four decades in manufacturing. Essentially, only 40 percent of labor is needed to do the same job as in the late 1980s. Not only is the effect enormous, it has also been rapid—more than a doubling in productivity in three decades. Of course, this has several consequences. It leads to more production both in terms of sheer quantity (say number of cars) and their quality (cars now have much more sophisticated electronics than they did three decades ago). But it also leads automobile manufacturers to require smaller workforces. This is not unique to manufacturing nor to recent times. For instance, gains in agricultural production have led the population of the U.S. involved in farming to decline from almost 70 percent in the early nineteenth century, to around 2 percent today, and not because humans are tightening their belts. At the same time, we saw dramatic increases in the quantity and quality of food produced.

Moreover, the explosive growth of production and exports in

Figure 6.5: This shows how many hours of labor are needed to manufacture the same output over time (measured in hours to produce the same GDP due to manufacturing, normalized relative to 1987). In 2016 it took less than 40 hours to produce what would have taken 100 hours in 1987. Data from the U.S. Bureau of Labor Statistics.

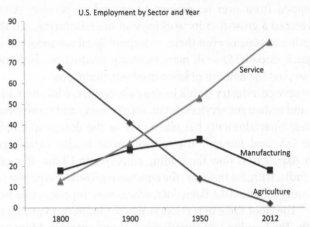

Figure 6.6: Agriculture has been a shrinking percentage of U.S. employment since the 1800s, and manufacturing began its slide in the 1950s. Although employment in services continues to grow, even parts of the service industry are seeing some drops in the amount of labor needed. Data from Piketty (2014), Table 2.4.

countries like China, Mexico, India, and South Korea has led to the internationalization of labor. As working classes in manufacturing and services emerged in much of Asia, South America, and parts of Africa, middle classes have shrunk in Europe and North America. It is important to note that most of that shrinkage was due to changing technology rather than globalization. For example, studies suggest that only between 10 and 20 percent of manufacturing jobs in the U.S. between 1999 and 2011 were lost to increased imports from China; most were lost to changes in technology.[17]

The combination of these forces is seen in Figure 6.6. The dramatic downward trend in agriculture is clear, and we are partway into the same trend in manufacturing. The drop in the labor used in manufacturing only started its decline after World War II, but is on an inevitable trend to tiny numbers too.

These trends are affecting not only the Americas and Europe, but Asia as well. For instance, as agriculture has been modernizing in India and China, similar trends have emerged in those countries. They are not yet down to having only 2 percent of their population being farmers, but they are halfway there. In China the fraction of the labor force working in agriculture dropped from 60 percent in 1990 to 28 percent in 2015, and in India over the same time period it dropped from over 60 percent to around 45 percent. Asia has experienced a growth in its workforce in manufacturing, but that is beginning to plateau even there, as technological advances are making quick inroads. Overall manufacturing production continues to increase, but the amount of labor used is flattening out.

The service industry (think health care, finance, education, media, hotel and restaurant services, retail, warehousing and transportation services, entertainment) has taken over as the dominant employer in the U.S. and Europe. The service sector is also rapidly growing in Asia, and is now the leading employer in China. It is taking off in India, with, for instance, the enormous growth of computer programming in cities like Bangalore, which now supplies code worldwide.[18] The labor force in services is also not immune to productivity growth. The London taxi situation is just one example. More generally, we see humans being replaced by computers and increasingly sophisticated programs: for filing taxes, teaching foreign languages, and trading financial securities, just to name a few.

These forces have led to widening gaps in wages, a trend that will only continue. In the U.S. in the 1950s, a typical worker with a college degree earned 50 percent more than a worker with only a high school degree or less. Now that difference is on the order of 100 percent: the earnings for college graduates have been increasing, while the earnings for those with only a high school degree are falling.[19] People with the highest levels of education are finding their skills amplified by modern technologies—they can make greater discoveries, reach more people, and perform tasks that were previously impossible. In stark contrast, people with middle to lower levels of skills are being replaced and displaced by technology.

The productivity explosion has not eliminated the need for low-skilled workers. Demand for theme parks has grown; the fitness and tourist industries are expanding, as are others that provide leisure goods. That which people used to do for themselves, such as cooking and cleaning, is now being done by low-skilled services. So we see increased demand for labor at the extremes—for design and management tasks that require high levels of skills, and for doing mundane tasks that require low levels of education and experience.

These changes have resulted in the percentage of people who end up with income higher than their parents' to fall almost in half since the Second World War. Figure 6.7 shows that more than 90 percent

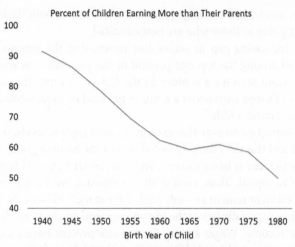

Percent of Children Earning More than Their Parents

**Figure 6.7: Comparing child's to parent's household pretax income measured approximately at age thirty and adjusted for inflation. Data from Chetty, Grusky, Hell, Hendren, Manduca, and Narang (2016a).**

of children born in 1940 earned more than their parents did—there was a booming middle class and growing prosperity and incomes for almost all Americans were rising from one generation to the next. Of those born in 1980 only 50 percent would earn more than their parents. The numbers are even worse when looking just at males, as nearly all born in 1940 were doing better than their parents at age thirty, but of those born in 1984, only 41 percent were—so 59 percent were doing worse than their fathers. The gains in overall productivity are being spread unequally: people in the upper-income brackets are more often than not doing better than their parents; people in the lower brackets are doing worse, even with education levels similar to those of their parents.

The composition of the labor force has not changed as rapidly as productivity. Thus, continued improvements in technology have led to a shortage at the very top where wages and productivity are highest, and a glut of labor where it is least valuable and wages are lowest. The percentage of twenty-five- to twenty-nine-year-olds who have a bachelor's degree rose from just over 20 percent in 1974 to just under 30 percent by 2014. This might seem like a big increase, an almost 50 percent gain. However, this is far below the change in relative productivity. The slow adjustment of the population to required education has accelerated the wage premium of education, with the largest benefits going to those who are best educated.

This increasing gap in wages also means that the income concentrated among the top one percent of the population is reaching extremes not seen for a century in the U.S. or Europe. The top *one percent* in Europe earn about a tenth of the total income, while in the U.S. it is around a fifth.[20]

It is important to note that even at the very highest levels, in both Europe and the U.S., roughly two thirds of the income going to the top one percent is labor income, and only about a third is from the returns to capital. Thus, even at these extremes, wealth advantages and investment returns are only part of the story.[21] Most of the rising inequality, especially below the top one percent, results from changes in labor income. Wages among the top one percent have increased more than two and one half times since the early 1970s; among the top 5 percent they have doubled; among the top 10 percent they have grown more than four and a half times. But among the bottom 60 percent, wages have gone up by less than a third over the same period. The rise in inequality is due to extensive changes in the relative wages of large segments of the population.[22]

With growing gaps in income between those who go to college and those who don't, an obvious and central question emerges: why aren't more people getting higher education?

This circles back to immobility: there are strong social forces that make it more or less difficult for different people to acquire the skills that they need to succeed in the modern labor market.

## Education

In the U.S., children with well-to-do parents are more than two and a half times more likely to graduate from college than children from poor families.[23] There are also corresponding differences across ethnicities. The percentage of people in their late twenties who have a college degree is an astonishing 72 percent among Asians, 54 percent among whites, 31 percent among African Americans, and 27 among Hispanics.[24] The causal mechanisms for these differences are intricate, as family income, ethnicity, parents' education, culture, and community are all correlated.

Beyond which children go to college, there are also large differences in which colleges they attend. If you sit down in an introductory lecture in biology or economics or computer science in one of the most competitive universities in the U.S. (e.g., Ivy League schools, Stanford, MIT), the student sitting next to you is more than *twenty times* more likely to be from the wealthiest quarter of the population than from the poorest quarter of the population.[25] Someone from the wealthiest one percent in the U.S. is 77 times more likely to go to an Ivy League school than someone from the poorest 20 percent.[26] The students going to top universities are still overwhelmingly from the upper income brackets, despite substantial financial aid that can make those universities an excellent and affordable investment for students from low-income families.

This has consequences. Not surprisingly to anyone who has searched for a job, the college one attends makes a difference. The U.S. Department of Education has a scorecard by which it values the returns to colleges. The median salaries of graduates from Harvard, MIT, and Stanford are more than twice the median college graduate salary.[27]

The eventual gaps in education are in play much earlier than college—and are already quite pronounced when children reach their teens. This is seen in Figure 6.8.

These differences begin just after childbirth. For instance, by the time they enter school, children of high-income families have heard an estimated thirty million words more than children of families on welfare. To arrive at this number, Betty Hart and Todd Risley[28] did something incredibly simple. They counted how many words parents spoke to their children. Parents on welfare spoke about 600 words per hour to their babies, while professional parents spoke more than 2,000 words per hour to their babies—more than three times more than the parents on welfare. This has an impact. Between 86 and 98 percent of the words that the children used were derived from their parents' vocabularies—and they have very similar speech patterns to their caregivers. Hart and Risley also found that language skills at age three correlated significantly with language and reading skills at ages of nine and ten.[29]

Early interaction snowballs. The benefits of helping children to

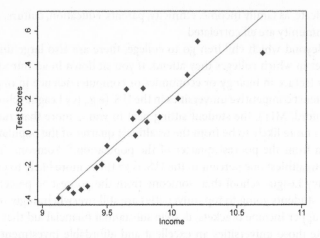

**Figure 6.8: The difference in average test score from the poorest to the richest neighborhood is a full standard deviation (so moving from the bottom third of the distribution to the top third). The test scores are the sum of math and English language arts achievement scores for eighth-graders (typically thirteen- to fourteen-year-olds) in New York City, in standard deviations. The income for a neighborhood is measured in log of per capita income for the neighborhood (based on an average using the zip codes containing the schools). There are twenty observations binned to be of equal size scaled by income. Data from Fryer Jr. and Katz (2013).**

develop at a very early age—in their preschool years—are highly leveraged. Teaching them things like patience and tenacity helps them learn other skills and opens doors for them throughout their lives. Teach a child to read earlier and they can start acquiring more knowledge at an earlier age, and this process accelerates. An improved ability to express themselves enables children to better navigate the world and learn more. The more they know, the better they perform in school and the more rewarding they find it, leading them to devote more effort to it. James Heckman, a Nobel Prize–winning economist and expert on early childhood education, put it this way: "skills beget skills in a complementary and dynamic way. Efforts should focus on the first years for the greatest efficiency and effectiveness. The best investment is in quality early childhood development from birth to five for disadvantaged children and their families"[30]

Some differences in parenting are due to time constraints, but others can be traced to differences in beliefs of the parents as to how valuable it is to spend time with their children. For example, in a recent study in the U.K., parents in the lowest quarter of the income distribution answer very differently from those in the top quarter of the income distribution when asked whether they agree with the statement that: "My child develops at his/her own pace and there is little that I can do about it".[31] Almost 40 percent of the poorer parents agree or are unsure about that statement, while less than 20 percent of the wealthier parents agree or are unsure. So, twice as many of the poorer parents are not sure whether they have much to do with their child's development. One might attribute the beliefs of poorer parents to rationalizing the lower amount of time they spend with their children. But other questions suggest it involves more. When asked about the statement "My child is capable of learning to programme a software," almost 80 percent of the richest parents thought it very likely, while only 50 percent of the poorest parents thought so.[32]

The role of information in education extends well beyond the early formative years.

Despite all of the disadvantages that a child has growing up in a low-income family and community, the poorest people in the U.S. still manage to produce many high-achieving students by the end of high school: those with a high grade average (at least A–) and in at least the 90th percentile on college entrance exams (SAT or ACT). Seventeen percent of high-achieving students come from families in the bottom income quartile, which is half as many as come from the top income quartile (34 percent).[33]

When all the effects of childhood are accumulated, the upper-income families have a two-to-one advantage over lower-income families in producing high-achieving children. That is enormous, but there is still a huge missing piece. Recalling our earlier numbers on education: there was more than a *twenty-to-one* ratio of upper-income compared to lower-income children at the top universities in the U.S.[34] So there is still a factor of ten difference by income level as to whether high-achieving students end up in a top university.

You might think that top universities are too expensive for the low-income students. This seems to be at most a minor factor. It is estimated that fewer than 8 percent of families in the U.S. are constrained in college choices by income.[35]

Still, it is hard for people to know how much a college will end up costing them, just as the true cost of other large purchases are unclear. In many countries, you would be silly to walk into a car dealer and pay the sticker price for a car. Except for the hottest new models, the sticker price is the dealer's starting point for a negotiation. As more consumers have become savvy, and many ask to see the invoice listing the price that the dealer paid for a car, the industry has begun to inflate those dealer invoices to include profit margins in various disguises, thus making it harder to know what lowest price a dealer would really accept. With a bit of research, and some shopping around, one can often negotiate a price significantly below the sticker price.

Similarly, the prices quoted for universities are starting points—the very most one could possibly pay. However, luckily, most universities are not as aggressive as car salesmen. For universities the difference between the starting prices and what most people pay is much more dramatic than the discounts with cars, and there is less haggling over the price. There are a variety of grants, fellowships, scholarships, and other programs that significantly reduce the price to most students. Only *one third* of full-time students pay the sticker price. In 2016, the average quoted price (inclusive of room and board) for a four-year private college in the U.S. was $44,000, but the average paid was only $26,000. Looking at public four-year universities in the U.S., the average net tuition and fees paid by the highest quartile is $6,330, while by the lowest quartile it is $-2,320. That is a minus sign in front of the last figure—the poorest quarter of the population receive payments that more than cover tuition and fees and also help pay costs of room and board. For private (nonprofit) four-year colleges, the richest quartile's average cost in tuition and fees is $19,720, while it is $4,970 for the poorest quartile.[36]

Even though the prices can be very low, many high school students and their families have little understanding of how it all works. They are either scared away from what they think the prices are, or they don't even know that the opportunities for substantial financial aid exist. Again, a lack of knowledge keeps low-income students from even applying to schools that might be the best matches for them.

Rejoining Claire Vaye Watkins's story gives us more insight here. As she states:

Most parents like mine, who had never gone to college, were either intimidated or oblivious (and sometimes outright hostile) to the intricacies of college admissions and financial aid. I had no idea what I was doing when I applied. Once, I'd heard a volleyball coach mention paying off her student loans, and this led me to assume that college was like a restaurant you paid when you were done. When I realized I needed my mom's and my stepfather's income information and tax documents, they refused to give them to me. They were, I think, ashamed. [37]

## *The Usual Suspects: Institutions and Capital*

Why should our well-being be so tied to that of our parents and the community into which we are born? This is by no means a new question, but one to which networks offer important insights. As we have just seen—information is key to understanding enormous differences in education and mobility. That information comes largely from the community within which the parents and children reside.

A notion of "capital" helps capture the information that one gets from one's community. But we need a notion of capital that extends well beyond its classic conceptions. Classic forms of capital—such as "financial capital," referring to financial assets that can be converted into other forms of capital or labor, and "physical capital," referring to material resources, including land—play minor roles in our story.

Alfred Marshall was already stretching beyond the classic concepts to include "human capital" more than a century ago. Marshall stated, "The most valuable of all capital is that invested in human beings; and of that capital the most precious part is the result of the care and influence of the mother." The more modern notion of "human capital" is the knowledge and skills that are used in any aspect of production and trade—including art, invention, design, operations, management, and marketing. Much of human capital, other than innate talent, is acquired: one learns how to do something—in school, by observing, as an apprentice, or by trial and error and experience.

The most relevant form of capital from our perspective is "social capital," which is also the most recent concept of capital. [38] It refers to the favors, resources, and information that a person can access from their network of social connections or as a result of their reputation. [39] The social connections providing social capital include friendships, professional relationships, as well as memberships in or access to organizations, both private and public. This is the broadest form of capital, the squishiest in terms of definitions, and it involves things that are not easily measured, such as relationships and position in a network.[40] For instance, social capital played a major role in our story of the rise of the Medici. It was not just that the Medici could call upon many allies for favors, but also that the Medici's unique position enabled them to coordinate others—which ultimately was a source of great power. The power that comes from that ability to coordinate others is also a form of capital as it can be very productive and valuable for the person wielding it.[41] Social capital relies on a variety of such concepts that are open to interpretation, such as reputation, position, friendship, and organization. Even homophily plays a role: a person can have a dependable and extensive network of connections and still lack access to vital resources and knowledge. Despite the difficulty of having a single, succinct, and easy to measure notion of social capital, it is such an important concept that we cannot avoid it.

As various forms of capital can often be converted into each other, having an abundance of one can enable one to acquire others. For instance, financial capital can be used to buy education and thereby human capital; and social capital leads to knowledge and opportunities that help build human capital and financial capital.

The fact that all the forms of capital are heritable—they can be passed from parents to children—has implications for immobility. Parents can give their children money, assets, and property. Parents are also their children's teachers throughout life—they are a child's first source of human capital, and play key roles in determining where children go to school and what and how they learn. Finally, and not to be underestimated, children are born into a particular family and community that form the foundations of their social capital. Moreover, the parents' own social capital also makes a difference in how they act toward their children, for instance how aware they are of the value of education and what they can do to help their children excel. Homophily, along lines of geography, income, ethnicity, and culture, means that these communities are often separated from each other and can be quite insular—limiting information flow and locking communities into like-minded behaviors.

Communities are very clumped—so that wealthy and well-educated live together and poor and poorly educated live together. Fewer than 10 percent of people living in East St. Louis, Illinois, or Benton Harbor, Michigan, have a college degree, while more than 80 percent of adults do in Upper Montclair, New Jersey, as well as in Palo Alto, California. In Pahrump, Nevada, Claire's hometown, 13 percent of the adult population over twenty-five years old have a bachelor's degree.[42] If fewer than one in seven adults has a college degree, it is not surprising that many parents and students in the community know little about how to apply to college or how it can be paid for. Moreover, if most of those who did go to college went to a local community college or a noncompetitive school, then high-achieving students like Claire will lack network connections to people with knowledge of the opportunities that exist at top universities.

Communities are highly segregated by income and only getting more so as income inequality increases. In 1970, almost two out of three people lived in neighborhoods in which the median income differed by no more than one fifth from the U.S. median income. By 2009, that fraction had dropped to just over two out of five, and so a majority of people live in a neighborhood with a median income that is substantially different from the U.S. median.[43]

In 1992 the U.S. Congress authorized an experiment leading to a major study of how neighborhood affects a family's well-being. The centerpiece of the study was called Moving to Opportunity for Fair Housing. Four thousand six hundred families living in public housing around the U.S. (in Baltimore, Boston, Chicago, Los Angeles, and New York) were selected to participate. They were randomly assigned to three different groups. Some were given vouchers to help pay for rental housing—provided they used it in a low-poverty neighborhood—so this group had to move to a wealthier neighborhood in order to use their vouchers. Others were given vouchers that they could use anywhere they liked, so they could stay in the same neighborhood if they chose to. The remaining families were not given any vouchers and so were a control group who were observed for comparison.

The Moving to Opportunity program took place between 1994 and 1998, and the children involved are now adults and so we can see how their lives were changed. The effects are dramatic.[44] The biggest effects were for the children who were youngest when their families moved. Raj Chetty, Nathan Hendren, and Larry Katz combined information about who moved and who stayed behind from Moving to Opportunity with later tax data from the IRS to see how where a child grows up impacts his or her income and life. If the children were under thirteen years old when they moved to a low-poverty neighborhood, their earnings in their mid-twenties were almost one third higher than those of the control group, who were not given vouchers. The estimated benefit for an eight-year-old child of moving to a low-poverty neighborhood via such a voucher is *three hundred thousand dollars* in lifetime earnings. The children who moved were also one sixth more likely to go to college, and the colleges they attended were significantly higher ranked; and they were less likely to live in poor neighborhoods or be single parents (at the time of the birth of any children).

You might think that the luckiest families were the ones that could use vouchers anywhere they liked, but many only used the vouchers to save rent and did not move. The extra money was useful, but did not have as big an effect on the children's lives as moving. The biggest gains were achieved by families who were required to move to low-poverty neighborhoods to use their vouchers. The income effect was about half for the group who were given vouchers but not required to move, and much of their gains actually came from those who did choose to move. Most notably, the effects on children are dramatically higher the younger they were when they moved.[45]

The Moving to Opportunity study confirms, in a very stark way, what many social scientists and others have been suggesting for decades: neighborhoods and communities matter substantially.[46]

## Job Networks and Social Capital

Understanding social capital in terms of access to job opportunities also illuminates immobility. It helps explain why, even after graduating from the same college with the same major, children from low- and high-income backgrounds have different earning potentials. A recent study from the U.K. provides some orders of magnitude.[47] Without accounting for which colleges the students go to, there is a 25 percent difference in median earnings between graduates from high- and low-income families. Once one compares students who attend the same college and study the same subject, this drops to 10 percent. So, about three fifths (15 of the 25 percent) of the difference in outcomes based on family background comes from how that background influences which college students end up in and what they study, and then the remaining two fifths (10 of the 25 percent) comes from how family background determines what happens after students exit school.[48]

The key to understanding the persistence of this difference is an appreciation of how important networks and social capital are in determining who gets which job and what pay.

"Among corporate recruiters, random applicants from Internet job sites are sometimes referred to as 'Homers,' after the lackadaisical, doughnut-eating Homer Simpson. The most desirable candidates, nicknamed 'purple squirrels' because they are so elusive, usually come recommended." So said a human resources consultant.[49] Your chance of landing a job of any kind without some sort of connection is low.

Want a job at a bakery? Here's a baker's thought on the matter: "We'll give anyone who walks in off the street an application. Likelihood of their being hired is slim, because we don't know them. . . . We don't tend to just take people off the street because I've had a lot of bad experiences. That's true of all ethnic groups including white Americans . . . it doesn't matter."[50]

Perhaps you just want to sort things at a recycling center. It should be easy to get that job without a connection. Think again: "We hire almost entirely by word of mouth. I have a couple of guys who brought in almost everybody, all the unskilled anyway. They bring in their friend, their cousin, their uncle, whatever. I have had maybe five guys come in off the street and only two or three have worked out even for a short time. We have a formal application, but it's really only for record keeping purposes."[51]

If you have ever struggled to find a job, you are not alone. Without well-connected friends or family it is hard to make it in *any industry*. The first detailed study of how people find jobs, published in 1951, was by George Shultz.[52] More than half of the textile workers he interviewed had found their jobs via friends.[53] As this might be special to the textile industry, Shultz also examined how workers in a variety of professions found their jobs.[54] For a wide range of jobs, secretaries, janitors, fork-lift operators, truck drivers, electricians, and many others, between 50 to more than 70 percent of jobs were found via friends.

Later studies confirmed that high percentages of jobs are found via friends and acquaintances throughout the spectrum of professions, from low-skilled jobs to managerial positions, and around the world. Finding a job without having some personal connection to someone already employed at a firm is more the exception than the rule.[55]

Like it or not, your fate is closely connected to that of your friends. If they are well-situated, then they can help you. If they are unemployed, then you are out of luck. About fifteen years ago, Toni Calvó-Armengol and I began studying the impact of this fact. What does the fact that job information is networked mean for patterns of employment and wages, as well as decisions to seek education and to stay in the labor market?[56] Having more friends who are well-employed provides advantages in a number of obvious ways. It increases the chances that one of them will hear about a job opening and pass it along, which in turn increases the chances that you will find a job that is a good match.

To get a feel for how this serendipity works, consider a simple scenario. Suppose that each job interview you get has a fifty-fifty chance of being a good match and leading to a job offer. For the purposes of our illustration, let's suppose if one does not work out, then the chance of the next one panning out is still fifty-fifty. If you get only one interview, then you have a one-half chance of landing a job. If you have two interviews, then your chance of landing a job goes up to three quarters—you only end up empty-handed if both opportunities don't work out (which happens with a one half times one half, or one-quarter chance). If you have three interviews it goes up to seven eighths, and with four you are up to fifteen sixteenths. The additional effect of each cumulative interview opportunity is half as much as the previous—so there are diminishing returns and eventually additional interviews do not make a big difference. But the first few have a huge impact.

The effect is not just on the chances of landing a job, but also on the wage. Let's suppose that if you get more than one job offer you can take the one with the higher pay. Suppose half of the jobs you might land are ones that you are overqualified for and pay only $15 an hour and the other half match well with your skills and pay $20 per hour. If you end up with one job offer, your expected wage would be the average, $17.50. If you get two offers, then you have a three-quarters chance that one of them will pay $20 per hour. With three offers, that probability goes up to seven eighths, and so forth. More friends lead to higher chances not only of employment, but also of a better job and higher wages.

It is important to note that one friend does not equal one interview, as not everyone you know is sitting on a lead to a job opening. Nor does one interview translate into half a chance at a job. This suggests that the network is more important than in the simple calculations above. Landing a job might require dozens of contacts who are willing to help you if they can, and landing a good match with a high salary might take even more.

This should come as no surprise—the more connections you have and interviews you land, the higher the chance that you end up employed and the higher your expected wage.[57] Having more employed friends yields a higher chance of landing a good job, or getting a better job if you are unhappy in your current position.

This has broader implications. For instance, friends' employment ends up being correlated. Suppose we examine two different groups. Among one group there is high employment, and so when someone becomes unemployed they have many connections who can lead to potential interviews. Among the other group there is low employment, and when someone becomes unemployed they do not have many fruitful contacts. People in the first group will have higher chances of landing a job when they become unemployed, and will also have higher average wages. As Toni Calvó-Armengol and I found, as one group becomes more employed and the other less employed, the former gets a greater advantage and the gap grows. This dynamic trends away from equality. When information about jobs comes through a network, friends' employment correlates: groups of friends in the aggregate tend to have either high employment or low—so feedback pushes groups toward the extremes.[58]

Although this makes good sense, and turns out to be true, it is actually tricky to establish that your employment is correlated with that of your friends *because* job information is networked. Your employment being correlated with your friends' employment is not convincing evidence that it is because of a network effect. Economists are skeptics—they want real proof. Why wouldn't such a correlation be proof? Your friends are not a random draw from the population. For instance, it could be that hardworking and reliable people tend to be friends with each other, and less reliable and lazier people also tend to be friends with each other—a form of homophily that would also lead people who are friends with each other to have correlated wages and employment.

If we could measure how friendships impact employment when they are randomly formed, we would glean real evidence that the connections actually cause friends to have correlated employment and wages. But where can such randomly formed friendships be found? Ron Laschever, an economics historian, had a great idea: he examined a military draft, which assigns people into relatively small groups for extended periods of time and strong friendships develop.

When the United States entered the First World War, it relied heavily on the draft to fill out its army. In the spring of 1917, the U.S. Army consisted of just under 300,000 men. By late 1918 it had over four million men, almost three million of whom were drafted. Men who were drafted were randomly assigned to one-hundred-soldier groups called companies. Over the span of the two years that many served, these doughboys would spend all of their time with their company. [59] They trained together, traveled together, fought together, and risked dying together. The bonds that they formed were strong and lasting. A decade later their reliance on their company mates was still strong. As Laschever found, a 10 percent increase in the employment rate among a former soldier's company mates in 1930 corresponded to a 4 percent increase in the chance that the soldier was employed. That is an impressive effect: it's less than one for one, but it is still a 40 percent ratio. Moreover, this effect is just from the employment of the former soldier's company, not even accounting for his connections to family and other friends. To be sure that it is really the connections to company mates and not something else about the military or the background of the soldiers, it is important to note that this relationship holds only relative to the soldier's own company. A soldier's employment does not vary with the employment rates of other companies that were formed around the same time and with similar demographic characteristics. So the soldier's employment is substantially dependent on the employment of his friends.

The military is not the only organization that randomly assigns people into groups. Students entering universities are assigned to dormitories. David Marmaros and Bruce Sacerdote took advantage of the random assignment of freshmen at Dartmouth College to see how their eventual employment fates, four years later, were intertwined.[60] The study looked at not only whether people were employed, but also at their salaries. Upon graduating, a student's employment was correlated with the fraction of their hall mates (students in rooms in the same hall in their dormitory) from freshman year who were employed: changing their hall mates from unemployed to employed led to a 24 percent increase in the chance that a student was employed compared to the average student. Moreover, each extra dollar that those hall mates earned led to a 26-cent increase in the students' earnings, above that of the typical student. So, changes in hall mates' outcomes corresponded to about one fourth as large a change in a student's employment and salary.

These effects are really lower bounds on how important networks are in employment, as they look at only particular groups of friends—Army buddies and college friends. As we age, the full list of all of our former acquaintances becomes enormous—there are thousands of people whom we end up knowing in some way. It turns out that even casual acquaintances matter. The person who sat next to you in history class in school, and just happened to bump into you last month and hear that you are between jobs, could be the person who puts you in touch with your new employer. A small number of our relationships remain strong ones—people we know well, can rely on for help in most any circumstance, and interact with frequently. Other relationships fall into many categories: childhood friends with whom we have occasional contact or could reach out to if needed; colleagues and acquaintances we have spent some time with but interact with only infrequently; distant relatives; friends of friends. . . . Many of these are weak relationships—people we know and could potentially contact for information or a small favor, but with whom we are in contact irregularly, or whom we see regularly but in a limited context.

This distinction between strong and weak relationships was the centerpiece of one of the most influential studies in the social sciences in the last half century.[61] Mark Granovetter examined the extent to which information about jobs flowed through weak versus strong ties among a group of people in Amherst, Massachusetts.[62] Granovetter found that only a sixth of jobs that came via the network were from strong ties, with the rest coming via medium or weak ties; and with more than a quarter coming via weak ties.

We have already seen that strong ties can be more homophilistic (comparing Figures 5.3 and 5.4).[63] Our closest friends are often those who are most like us, and live close to us, work with us, or study with us. This means that they might have information that is most relevant to us, given our shared backgrounds and interests, but it also means that it is information to which we may already be exposed. In contrast, our weaker relationships are often with people who are more distant both geographically and demographically. Their information is less redundant. Even though we talk to these people less frequently, we have so many weak ties that they end up being a sizable source of information, especially of information to which we don't otherwise have access.

Although any single strong tie can still be more influential than a weak tie,[64] casual acquaintances and former colleagues still make a big difference. Marie Lalanne and Paul Seabright[65] traced the pay packages of 22,389 top executives of 5,064 European and U.S. companies to their networks of connections. They counted how many former colleagues a given person has who are now in influential positions. If you once worked together with John and Alison, and now Alison is a top executive at some company and John is unemployed, then Alison counts as a connection but John does not. Some of these are strong ties, but many are weak—just people you happened to work with at the same company at one time. And these connections are only part of a person's network: former co-workers, so it will underestimate network effects. But these connections do capture people who are in the same profession and most likely to have relevant job information. The typical (median) executive in this network has more than sixty connections, and the best-connected executives have several hundred connections or more. What Lalanne and Seabright show is that if you compare an executive with the median number of connections to one who is in the 75th percentile, all else held constant—so, take an executive with around sixty connections and add a few dozen connections—then the executive with more connections has a salary that is 20 percent higher.

The correlation of connections with salary also turns out to be significantly higher for men than women: having more connections leads to a bigger increase in salary for men than women. This might help explain why men end up with wages more than a fourth higher than women, despite women having slightly higher education levels on average.[66] This is consistent with other findings of women being disadvantaged when it comes to referrals. For example, Lori Beaman, Niall Keleher, and Jeremy Magruder, in a study in Malawi, found that men tend to recommend other men even if they know well-qualified women, while women's tendencies to refer other women are not enough to overcome the effect.[67]

How this can have a big impact is then easy to see. Consider a profession with more men than women. Both men's and women's former colleagues will tend to be predominately men. Gender homophily also means that men are more likely to stay in touch with each other, and women with each other. Even though over time men and women might have similar numbers of past colleagues, since most of their ties will be to men, given the homophily women will have fewer ties to people who actually refer them for a job when they have the chance. In higher management ranks, which are heavily male populated, this bias can have a big kick. As we saw in our simple calculations, having a few more opportunities can result in a sizable difference in the quality of the job and salary.

Why are employers so enamored with referrals? If they were hiring the best talent out of the full pool of whoever applied, that would eliminate many of the distortions that networks introduce into labor markets. There must be gems among a wider pool of applicants, so why limit oneself to hiring friends and acquaintances of current employees?

Just as homophily helps in navigating a network, it helps employers find employees with specific characteristics. Suppose that you run a business and you need to hire someone who can work nights and weekends; or you need someone who likes to travel and speaks Spanish; or perhaps you need a programmer with knowledge of certain kinds of databases. If you already have employees who are filling such roles, then homophily tells you that they are very likely to be friends with someone who is similar to them. Your current workers' friends are likely to be the sort of people you need to fill your opening. Who better to help you find a programmer with the skills to design certain types of software than your current programmers who are already working on related software? Not only might they know someone who is qualified, but they also have an excellent idea of exactly what the position requires. Random applicants to some advertisement that you post—the "Homers" described earlier—are much less likely to match well with the needs of the position.

When you call customer service with questions about your credit card, the person you are talking to is likely in the middle of a busy shift. You are one of roughly five thousand calls that the person on the other end of the line is handling a month. Finding and retaining the staff to handle such calls is a challenge. When a call center finds someone who works out, having them refer their friends is a good way to find others. Indeed, in one study, homophily was strong and people referred by current employees were similar to those employees in terms of gender, education, tenure in previous/current job, and prior/current wage.[68] It is then not surprising that those referred are almost twice as likely as other applicants to get a job (11.9 to 6.7 percent); and not only are they more likely to be interviewed, but more than 50 percent more likely to be hired conditional on being interviewed. Moreover, on average they were over $400 cheaper in terms of interviews and related costs to hire, but only cost a $250 bonus to the person who referred them.[69]

Referrals from employees, and other personal connections, also contain information about the reliability and other personal attributes of a potential applicant. Each year, when we hire new assistant professors in our economics department at Stanford, we end up with four or five hundred applications for one or two positions. Some are easy to sort through as they are missing some basic qualifications, but that generally still leaves hundreds of applications out of which we have to identify the few candidates who are most likely to be highly successful. Are they hardworking and creative? How well do they collaborate with others? How well will they fare as a teacher and mentor? You will rarely find answers to such questions in a résumé or even by reading things the applicant has written. Letters of reference are vital, but how can you trust what is written there?

To test whether referrals convey useful information about an applicant's potential performance that an employer could not predict from just looking at a résumé, Amanda Pallais and Emily Glassberg Sands did some hiring experiments via an online work site.[70] They found that referred workers performed better and stayed longer, beyond what you could predict from their résumés: the fact that an applicant was referred in and of itself meant that the person was likely to outperform someone hired who had similar traits but who was not referred.[71]

## Feedback and Persistent Differences in Behaviors: A Social Tug-of-War

The role of networks in providing job opportunities leads to more general complementarities in decisions. If many of my friends drop out of the labor force, then it becomes harder for me to find a good job, or one at all, and easier for me to find work illegally. Their dropping out can also signal to me that working illegally might be the best decision for me as well. My friends may even directly pressure me to follow their lead.

All of these forces—from parents, peers, and community—lead to cascades of behavior, just as we saw in our discussion of diffusion and contagion. The interesting twist comes when we couple these forces together with geographic and demographic homophily. The resulting contagion behaves very differently from the purer contagion that we saw earlier.

That contagious diseases could jump stark boundaries in social networks was particularly evident in feudal societies. Nobles lived separate lives from peasants, with strong differences in literacy, diet, and wealth; but when a disease spread it knew no boundaries. Even a few interactions between classes enabled a disease to become rampant among all classes. The list of monarchs who died alongside many of their subjects during the smallpox outbreaks in the eighteenth century is impressive: King Louis I of Spain (1724), Tsar Peter II of Russia (1730), Louise Hippolyte, Princess of Monaco (1731), King Louis XV of France (1774), Maximilian III Joseph, Elector of Bavaria (1777), Alfred, Prince of Great Britain (1782).[72]

In contrast, decisions such as whether one wants to go to college are influenced by *many* people around you and the outcome is more like a tug-of-war. It is not enough to meet just one person who went to college to know what it is about and how to get there and to become motivated to go. Many of our decisions end up being driven by that with which we are most familiar and who influences us the most. This differs from disease contagion, as it is enough to know one person with smallpox to catch it.

Thus, although many diseases easily jump across social boundaries, behaviors that depend on reinforcement and strength of interactions do not.

To see this most starkly, let us look at an example of a "dropout game" that is representative of some settings that Toni Calvó-Armengol and I examined.[73] We will keep the illustrative game simple: a person decides either to stay in school or to drop out (or to go to college or not, or to commit a crime or not, etc.). Most importantly, the person decides to drop out if more than half of that person's friends have dropped out. This is the social tug-of-war: one ends up pulled in the direction of most of one's friends. But to add the critical ingredient, homophily, let us suppose that there are two types of people: solids and blanks. Solids tend to be friends with solids and blanks with blanks, as pictured in Figure 6.9.

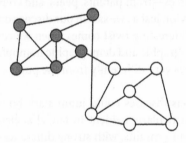

**Figure 6.9: Homophily and dropouts: A society in which there is homophily: a majority of the blanks' friends are blanks and a majority of the solids' friends are solid.**

Let us see what happens beginning with two dropouts among the blanks to seed the process. Figure 6.10 shows how this behavior cascades to infect all of the blanks, but stops short and does not infect any of the solids. Homophily acts like a firewall—behaviors are completely different on different parts of the network.

To see how homophily was necessary for the spread of dropout behavior among the blanks, note that if we rewire the network to have more cross-type relationships, the dropout behavior stops early. This is pictured in Figure 6.11.

Although this example is overly stylized, it is easy to see how

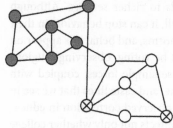

(a) The blanks start with two dropouts.

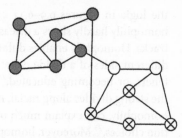

(b) A majority of the friends of the two middle blanks have dropped out, and so they drop out too.

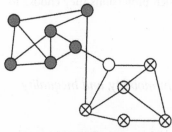

(c) Now, a majority of the top-right blank's friends have now dropped out, and so she drops out too.

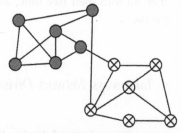

(d) Finally, the last blank has most of his friends drop out and so drops out. The dropout effect stops here due to homophily.

**Figure 6.10: Cascading dropout decisions, starting with two dropouts. Each person does what a majority of his or her friends do: if the majority of a person's friends have dropped out then he or she drops out too; if a majority are still in, then the person stays in.**

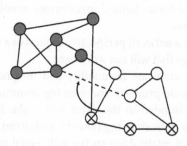

**Figure 6.11: Homophily and dropouts: Reducing the homophily can stop the spread of the behavior. In this example, changing one key link between two blanks to instead connect a blank to a solid stops the spread of the behavior.**

the logic in Figures 6.9–6.11 extends to richer settings. Although homophily hardly slows a disease at all, it can stop behaviors in their tracks. Homophily enables different norms, and behaviors appear on different sides of a divide, whether it be voting, observing religious tenets, or becoming educated.[74] These simple forces, coupled with the strong divides along racial, income, and caste lines that we see in homophily, can explain much of the observed correlation in education choices.[75] Moreover, homophily affects not only whether college is in a student's future, but also many related behaviors such as how much time they spend studying and paying attention in class, what they do with their free time, and which professions they choose to pursue.

## Takeaways: Network Divisions, Immobility, and Inequality

Inequality and immobility are like social cancers. One can treat the symptoms through taxation or income redistribution, but although these actions can be ameliorative they do not cure the disease. Fortunately, immobility and inequality are easier to diagnose than many cancers. But just as with cancer, many factors, often interacting, lead to immobility and inequality. Understanding what causes them helps us to prescribe a cure. Entrenched forces of homophily, coupled with the information and opportunities that flow through one's network, are fundamental forces behind investments in education, immobility, and inequality.

So what does a network perspective suggest as a cure? Let's start by ruling out things that will not work here.

One way to eradicate immobility and inequality is radical: by eliminating private property and forming communes, such as Israel's kibbutzim founded in the 1930s and 1940s. My colleague Ran Abramitzky has been studying how the kibbutzim have evolved and the challenges of establishing such a fully equal society.[76] Although the kibbutzim do have some range in ideologies, many of the typical founders and early members were Jews from Eastern Europe who had strong Marxist tendencies and full equality in mind. There was no private ownership of property, all income was shared collectively, and members ate in communal dining halls. Even children were raised communally, living separately from their parents in order to free women from disproportionate burdens of child-rearing. Children were taught altruism, cooperation, the importance of the collective, and socialism. At the peak of the kibbutzim, in the first decades after Israel became a state, over 5 percent of the country's population lived in a kibbutz. Many of the kibbutzim were located in key locations that helped secure the Israeli state, and they provided much needed safety nets not only as people were forming new settlements, but also a new country. For example, substantial numbers of new members fell ill with malaria and were sustained by the collective as they adapted to their new environment. In the 1950s and 1960s, the kibbutzim were some of the most productive parts of the Israeli economy.

By the 1980s, however, the kibbutzim were experiencing steady decline in relative population. By the year 2000 only 2.5 percent of Israel's population lived in a kibbutz. They also experienced a brain drain: the most educated members were exiting at a much higher rate than well-educated members were joining. New membership was coming mostly from those born in the kibbutz and newcomers with low education levels. The work ethic in some of the kibbutzim also sagged. The incentives to participate and work hard were muted in a fully egalitarian system that relied entirely on the ideology and altruism of the members. As the most productive members began leaving or working less, some kibbutzim faced financial crises. They were forced to back away from full egalitarianism; many kibbutzim instituted policies by which members were allowed to keep some of their income and could hold private property; collective sharing was now only partial. Many even began contracting out for health and education services. By 2004 only 15 percent of the kibbutzim were full cooperatives.

The transformation of the kibbutzim was seen on larger scales in other communist systems, such as the former Soviet Union and China, which transitioned to market economies, due to not only incentive problems, but also the logistical challenges of running massive planned economies. What was clear was that there are trade-offs among equality, incentives, and productivity. The huge advantage of market-based economies in allocating goods and services and providing incentives for innovation and growth means that most of us will have to tackle inequality and immobility by means other than the elimination of private property and the institution of communes.

In so doing, we are faced with some inexorable trends. Advances in technology will continue to displace low- and middle-skilled labor and increase the power and leverage of high levels of education and skills, pushing the remaining workforce into jobs that require little special training and are increasingly mundane. In addition, in so far as markets are global, over the next few decades more of the world's population will be pulled out of poverty and more will become better educated. To anyone paying attention, the growth of Chinese universities in both quality and scale is impressive. Dramatic demographic changes in one country, such as increasing labor supplies at all skill levels in India and China, have international ramifications. Acknowledging these facts leads to several observations and prescriptions.

Governments are always tempted to shore up the demand for labor in domestic industries with trade barriers and immigration policies. However, given that it is technology that is driving the displacement of lower-skilled labor, such policies will do little to stimulate employment, and might result in catastrophic collateral damage (as we will see in Chapter 9).[77]

As it is impossible (at least in the short run) to have everyone highly educated, this will lead to increasing polarization in the workforce and in the political sphere.

Past is prologue here. Throughout the nineteenth and early twentieth centuries, when farm labor was declining, compulsory education expanded, requiring that children attend school until a minimum age (between fourteen and eighteen years in most countries). This provided more literate populations and flexible workforces, better able to fit into the growth of more skilled blue- and white-collar jobs.[78] The current technological changes are pushing the returns on skills to much higher levels, suggesting that countries should be investing in ever higher levels of education to match technological advances. Looking forward to our *Star Trek* future, labor will be reallocated in ways that we can only imagine; skills and education will be needed in subjects that we cannot currently conceive of.

To deal with the income disparities between high- and low-skilled labor, some support at the lower levels of income is needed. The good news here is that the world is much more prosperous than ever before, and the enormous technological changes have led to a bounty that can be shared. Finland, among others, has been experimenting with what seems to be a radical approach. They are randomly selecting thousands of unemployed Finns and simply giving them money and benefits without caveat—these beneficiaries don't have to stay unemployed or prove that they are looking for work or enroll in any programs; they simply get the money. This idea of "universal basic income" has been gaining traction, and similar trials are on the horizon in Canada, India, Kenya, the Netherlands, and even a private venture in California.[79] It is an obvious, but highly controversial, policy in response to an ever-polarizing workforce.[80]

Regardless of the success or failure of universal basic income, such a policy still does not address the disparities in opportunities that underlie immobility, and which are both unjust and counterproductive. Much of this, in one form or another, traces back to the homophily that limits information and opportunity flows to both parents and children, leading to systematic underinvestment in education among lower-income households that have little experience with colleges or contact with highly educated people. We also saw that homophily leads to a variety of feedback effects. Limited opportunities and knowledge of those opportunities further limits the community's access and knowledge. Network feedback also leads a community to have correlated employment and wages, and also correlates the decisions of its members on things like investing in human capital, choosing a profession, and dropping out.

So, how do we counter the deleterious effects of homophily? Making dramatic changes in people's networks is likely a losing battle— one can try to raise awareness of homophily and steer networks in the right direction, but large-scale social engineering has a history of disasters; and one cannot move everyone to opportunity. Nonetheless, establishing more ties by providing short-term moves should still make a difference, especially as weak ties can be valuable.

The more immediate prescriptions are to undo what homophily does: to provide information and opportunities where they are lacking. Information is cheap and can be very effective. It is like preventive medicine, which saves enormous costs later. Helping people understand the importance of a few behaviors, and providing periodic but forceful nudges, can be remarkably successful.[81] Informing parents about the importance of early education and how to help their children learn, from preschool onward, is the easy starting point. The diffusion of information and behavior via networks means that feedback will help here—the more information is out there, the easier it is to get it to spread. In addition, removing obstacles that handicap lower-income working parents from giving children attention, such as providing affordable preschools and easy access to after-school programs, can also help. The impact of such programs is already visible: the percentage of four-year-olds in state-funded preschools has doubled since 2002, and the readiness gap between high and low incomes is correspondingly falling.[82]

The next level of information concerns opportunities for higher education and careers. As becomes clear from Claire's story, many high school students lack basic information about the world beyond their community. Informing students about the value of an education has been found to help their attendance and performance, from the Dominican Republic to Madagascar.[83] Providing basic information—along with some training and oversight—also helps in recruiting low-income students to universities. Stanford student Jeffrey Valdespino Leal's parents did not graduate from high school. He had thought that colleges outside his state would be too expensive until he was invited to a workshop where he learned about financial aid. Then the American Talent Initiative financed by Bloomberg Philanthropies connected Jeffrey with a student at Williams College who helped him with his essays and applications. As Jeffrey states, "If there could be more lower-income students here, it would be great, because we've shown we can do just as well as the other students."[84] Simply providing information to kids in high school has its limitations, as just because a teenager is told what they need to do to improve their future does not mean that they will do what is necessary, especially if no one else around them is doing it.[85] The aid from a college student was likely instrumental in helping Jeffrey out. In addition, such help can be leveraged by network feedback: much like our dropout game, people tend to follow the herd (more on this in Chapters 7 and 8). There is much more leverage in concentrating efforts and getting groups to change behavior than just providing information or helping a single student here and there. As more students are drawn from low-income communities, that will lead to further increases. The resulting increased mobility not only levels the playing field, but should lead to increased productivity.

In the end, given the growing premium to high-skilled labor and cost of inequality, it becomes clear that the countries with the brightest future will be those with the highest levels of education and the lowest hurdles to mobility. Technological changes disadvantage countries with large fractions of workers who have no choice but to be low skilled, and such countries will be left behind.[86] Educating people from the full range of the income distribution becomes vital, whether they come from the slums of Kolkata, rural Mongolia, or a poor neighborhood in Chicago. Increasing mobility not only addresses moral obligations to provide equal opportunities, but also lowers future health, welfare, and law enforcement costs, and increases economic productivity. It is essential for addressing widening inequality and for growing economies. One of the main insights that networks offer is how coupled decisions, opportunities, and outcomes are across network connections in a society. These complementarities mean that there are benefits from carefully coordinating policies, not to just randomly help individuals, but to take advantage of that complementarity. Helping one person each in two different communities can have less impact than helping two people in one community, which can then start to overcome thresholds to get others to act (and more on this in Chapter 8). Understanding the nature of the complementarities can lead to drastically different policies than just targeting categories of people without taking advantage of their network positions and influence.

# 7 · THE WISDOM AND FOLLY
## OF THE CROWD

What separates humans from other species? Many species exhibit complex social organizations, from the quadrillions[1] of ants whose elaborate colonies cover the globe to the spotted hyenas whose competitive matriarchal hierarchies make them the most fascinating species to roam the Serengeti. Humans are also not the only species to use tools. Jane Goodall dispelled that myth when she observed chimpanzees using blades of grass to fish delicious termites out of their mounds. Nor are we even the only species to teach things to our young. Meerkats teach their young pups how to handle scorpions by first bringing them dead ones, then later bringing them live ones with disabled stingers, and eventually bringing them healthy scorpions. Moreover, Meerkats cooperate in this endeavor: adults other than a pup's parents help to teach the young. Humans are not even the only species to communicate with sounds, as important forms of communication have been observed among elephants, whales, birds, and many other species.

If other species have elaborate social structures, cooperate in rearing and teaching their young, communicate, and use tools, then what sets humans apart? The answer lies in our ability to grasp abstract concepts combined with our ability to communicate them to others. I was not alive in seventeenth-century China, and yet I believe that peasant uprisings and rebellions occurred there. Moreover, I cannot ever be absolutely sure that this happened—all the evidence comes from artifacts, writings, and historical accounts and research. I can read about the period, and talk to experts who have studied that moment of Chinese history in detail, but I will never have any firsthand experience of the Qing dynasty. Nonetheless, through the variety of sources of information available I can learn a great deal about what happened and be fairly confident of the basic facts. It may be some time before I have an opportunity to travel into outer space, but I can read Michael Collins's account of Apollo 11's voyage and get an impression of what it must be like.

I have no idea of how to build a computer, much less the many tiny components that go into it, or of the minerals needed to build some of those. In fact, it would be impossible for any single human to produce even a few of its parts from scratch. Nonetheless, computers are produced by teams of people and chains of companies, and I can learn enough about a computer to use one to perform the research and writing needed to study networks and create this book.

Ken Mattingly was an astronaut of Apollo 16, and one of only two dozen people who have flown to the moon. In his words, "This is such a big thing. I frankly don't see how you can do it. Even though I am participating in it, I think it's audacious to even try. I clearly could not understand, as a crewman, how to, how to make it work. I could only learn to operate my share of it."

The human ability to grasp abstract concepts, which enables us to learn from each other, and to coordinate our activities, is a double-edged sword. We learn falsehoods as well as facts. Do vaccines lead to autism in our children? Is climate change a result of man's activities? We cannot answer these questions based on our personal experiences, we have only anecdotes at best, and so we must rely on what we have heard from others and from trusted sources. This allows the dramatic and persistent polarization that we see in people's beliefs on questions of fact. It enables man to achieve great scientific and engineering advances, but also leaves us open to doubts, superstitions, and polarization.[2]

In this chapter we will explore how we learn from each other and when we get things right and wrong. There are systematic errors that we make in interpreting information that we obtain from friends and acquaintances. For instance, we often treat similar information from different sources as being independent confirmation of a fact, even though it may emanate from a common (unreliable) source. Divisions in networks—and homophily in particular—can lead to persistent differences in beliefs and norms across groups. Our ability to communicate in the abstract makes us susceptible to deception, and it enables errors and even fake news to crowd out the absorption of facts and real news. Despite all of these challenges, there are situations in which we get things right. We will see how our networks determine when the crowd really is wise and when it is prone to folly.

## The Wisdom of the Crowd

Suppose that you need to estimate the weight of an ox, but don't have a scale. How would you do it? If you are asking yourself why on earth you would ever need to weigh an ox, it is likely that you will never have to. But don't worry, the example has an interesting history and point to it.

A standard method is as follows. You first take a cloth tape measure and measure the heart girth of the animal, in inches. This is like a chest measurement, measuring the circumference of the animal just behind its front legs. Next you measure the length of the animal. You measure from the point-of-shoulder to the point-of-rump (what is known as the pin bone). Basically, you are measuring from where the neck meets the front legs almost to the tail. You are advised not to sneak up on the animal, and to make sure it is calm and comfortable, especially when taking its heart girth measurement. Now you combine these numbers to estimate the volume of the ox—much like you calculated the volume of a cylinder in grade school. To estimate the weight in pounds you square the heart girth, multiply that by the length, and divide by 300.[3]

If you don't have a tape measure, but have a lot of friends, then another method is to ask each of your friends for a guess and then take either the median or average guess. The amazing accuracy with which this worked was noted in an article called "Vox Populi" published by Francis Galton in the scientific journal *Nature* in 1907, and brought to popular attention by James Surowiecki a century later.[4]

Vox Populi is Latin for "the voice of the people," and this phenomenon has become known as "the wisdom of the crowd." Sir Galton went to the annual show of the West of England Fat Stock and Poultry Exhibition held at Plymouth. There was an ox to be slaughtered and a contest was held. For a six-penny fee, you could venture a guess at the animal's weight. Those coming closest to the weight would win the contest. Eight hundred people entered the contest and Galton was able to read the guesses on 787 of the entries. The ox ended up weighing 1,198 pounds. The average guess was just one pound shy at

1,197 and the median guess was just 9 pounds over at 1,207—both within one percent of the actual weight![5]

It was likely a more ox-savvy crowd than might include you or me. In fact, half of the guesses were within plus or minus 3 percent of the actual weight and more than 90 percent of the guesses ranged between 1,000 and 1,300 pounds. Nonetheless, the striking aspect of Galton's analysis is the extent to which individual errors get washed out in the aggregate.

There are a few things that are critical to Galton's example working.

One is that there is a diversity of views. Almost eight hundred different people each drew on their own personal experiences to venture a guess. The diversity of those experiences produces a range of opinions from which we can learn.[6]

Next, the experiences and viewpoints cannot be biased in any systematic way. For instance, suppose that everyone used the same technique for estimating the weight of animals: for instance, they all used the tape-measure technique. That would do two things. It would reduce the diversity in their estimates, so that the differences in opinions would come from variations in measurements. That would not necessarily be a bad thing; however, it would also introduce a systematic bias. That measurement technique may systematically over- or underestimate weights.

For instance, suppose that the ox happens to be from a species that has more of their weight in their rears and less near their chests than other species. This would lead the tape-measure technique to underestimate the weight since this ox would have less weight in the place where the heart girth is measured. So, by all using the same technique to arrive at their guesses, they would tend to underestimate the weight of this ox on average. It would produce an interesting difference: instead of ranging from 1,000 to 1,300 pounds, it might be that everyone would guess somewhere between 1,130 and 1,180. Most people would not be as far off, but they would all tend to be wrong in the same direction—and so instead of having an accurate average, the average would be biased. Such systematic errors are a constant challenge for science: whenever people use the same technique or work from the same data they can be subject to common errors.

Finally, the various views have to be aggregated. If instead of picking the average or median, we had picked a guess out of a hat, or picked the maximum guess, we would not do as well as picking from the middle of the distribution. If done properly, aggregating the views of a group can outperform any individual, or at least a typical individual.[7]

An essential task in any organization is processing and aggregating information from multiple sources, both internal and external. There are many ways in which it is done, and so let us have a look at a few key ones.

For the latest views on a range of events from political elections to sporting events, one place to get some of the most accurate information is from prediction markets. For instance, to predict who will win an election, a prediction market enables people to bet on who will win instead of just answering a question in a poll. You can buy a share that pays one dollar if the candidate wins and nothing if he or she loses. If you think that there is a 60 percent chance that the candidate will win, then you can think of this share as being worth 60 cents in expectation: the 60 percent chance that it will pay a dollar. If a share is selling for less than 60 cents, then by buying it you expect a profit. If it is selling for more than 60 cents, then by selling it you expect a profit. The price then becomes a sort of tug-of-war between people who have different beliefs about the odds that the candidate will win the election—coming to rest at a point that balances the buying and selling pressures.[8] This is similar to the way in which betting on some sports events works.

Of course, getting accurate prices in a prediction market requires either the diverse and unbiased participation that we discussed above or someone who has very accurate knowledge, a lot of confidence, and a lot of money. Prediction markets offer several advantages compared to polls or some sort of averaging. One is that the market adjusts depending on how confident people are. If someone is very confident in their assessment, then they can buy or sell many shares and drive the price up or down accordingly. Another advantage is that prediction markets operate in real time and can adjust quickly to new information as it becomes available. A third advantage is that such markets allow the participants to see the current best estimate and that may help them hone their own thinking. This third feature can sometimes also be a disadvantage, as people may lose faith in their own estimates if they seem to be too much of an outlier compared to the market, even when their own estimates are good ones.

The accuracy of prediction markets in many elections (e.g., the Iowa Electronic Markets), and their ability to outperform many polls, has led to their adoption in a variety of arenas.[9] They have been used by many companies, including Google, France Telecom, Intel, HP, Eli Lilly, IBM, Microsoft, among others, to forecast things ranging from sales volume to interest rates. The U.S. Department of Defense even considered using prediction markets to help various intelligence and military personnel combine their information to forecast geopolitical trends and potential terrorist activities. That idea was scrapped as the public found "betting" on such events repugnant.[10] Tom Daschle, the Senate minority leader at the time, said, "I can't believe that anybody would seriously propose that we trade in death." California senator Barbara Boxer found that "there is something very sick about it," and suggested firing those who had proposed the idea. That does not mean that governments around the world will stop trying to forecast major events, but just that most will continue to find themselves unable to use markets to forecast certain events.

Another way to aggregate diverse views of what might happen, or what to do, is by having careful deliberation. This is the logic behind how juries work. It has also been used on larger scales, such as in what became known as Kasparov versus the World. In 1999, Garry Kasparov, one of the greatest chess players of all time, played a game against tens of thousands of enthusiasts who discussed strategies and voted on moves online. Kasparov played the white pieces and eventually won, but the match lasted four months and took sixty-two moves. Although the crowd did not win, Kasparov called it the greatest game in the history of chess. It is also clear that the crowd played at a level well above what most, if not all, of its members would have played individually. It also turned out that Kasparov had not only the advantage of playing the white pieces, but that he was also reading the discussion on the world forums.[11]

Most of our information, however, is not aggregated via some poll, vote, market, or deliberative process.

Suppose that you have to make a decision, such as whether to buy this book, whether to vaccinate your baby, whom to vote for in an upcoming election, whether to join a protest, or which type of phone to buy. In gathering information before making your decision, you don't have the luxury of averaging hundreds of people's estimates, or running a prediction market, or asking for a vote on what you should do. This is where social structure plays a major role. You are the aggregator and your network matters: you learn from talking with friends, family members, and colleagues, as well as from various media that you consult or follow.

Processing all of the information around us is tricky. When a friend recommends that you don't vaccinate your baby because they heard that it is dangerous, how do you know what to do? Where did their information come from? If you talk to more friends who say the same thing, is this new news? Could they be learning from the same source? Then you talk to another friend who says they just saw a news story saying that a new study found that vaccinations are safe and you should vaccinate your baby. How reliable was that news story and the study? Did your friend interpret the story correctly?

How will your opinions and beliefs evolve over time? Will you eventually accurately aggregate all the information that is out there just by interacting with your friends and acquaintances? How might you be fooled? How quickly do your beliefs adapt and change? Will you and your friends eventually reach a consensus? Is it possible for you to reach different conclusions than someone else in the same network? These are questions that economist Ben Golub (a former student of mine) and I investigated in detail.

To see how things can go right or wrong in a decentralized social setting, let's go back to our ox's weight. You have a guess but knowing little about the weight of oxen, you decide to talk with some of your friends. Some might have higher guesses than yours, pulling your impressions upward, and others might have lower guesses, pulling your impressions down. After talking with your friends your impression would be a "weighted" average (no pun intended) of your and your friends' initial guesses. The weighted aspect reflects that you might pay more attention to some friends than others, depending on how accurate you expect their opinions to be. For instance, if one of your friends is a rancher and the other is an economist, then the rancher's opinion might affect your view much more.

This sort of deliberation is very natural and could even optimally process the information—much in the way that Galton found the average of all the guesses to be very accurate. In your situation, however, the process does not end here. Your friends are also talking around, so that is changing their opinions too. Their new opinions reflect new information from other people in the network, including people who are not your friends. Thus, it is worth talking to your friends again. After the second time that you talk to your friends, you are incorporating information from friends of friends.

With a topic like the weight of an ox, you might get bored quickly and the process would end after a couple of iterations, at most. However, with topics more pertinent to your life, such as how likely a vaccine is to harm your child or how much you should be exercising for your health, you might continue to talk to people. Over time, your beliefs are incorporating information from friends of friends of friends.

This process is illustrated in Figure 7.1.

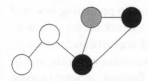

) The initial estimates, before any conversations. hter shades are lighter guesses. The intermediate shade of gray is the correct guess.

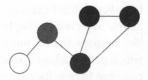

(b) After people have talked with each of their friends, they update their estimates. Their new estimates are an average of their initial estimates and those of their friends. Many of them are now some shade of gray. The person at the bottom left still has not changed beliefs at all, since she and her friend began with the same low estimate. The person who started with the correct estimate has actually gotten darker, now overestimating, since he is friends with two people who started with overestimates.

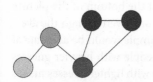

eople talk again and update their estimates again. t people are now moving to a middle shade of gray, the person at the bottom left is a very light shade of gray.

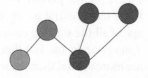

(d) The beliefs are getting ever closer to each other.

**Figure 7.1: Networked learning. Five people in a network are estimating the weight of an ox that really weighs 1,200 pounds. The two people toward the left, with the white nodes, start by underestimating the ox's weight and initially thinking it weighs 1,000 pounds. The two people with the black nodes start by overestimating and initially think it weighs 1,400 pounds. The person with the gray fill starts with a very good estimate of 1,200 pounds. So, lighter fill indicates a lighter estimate of the ox's weight. People talk repeatedly with their friends, and each time, they take a new average of their latest beliefs and their friends' latest beliefs.**

The process in Figure 7.1 shows that learning via a network has some similarities to diffusion and contagion. If we are each talking to a few friends, this ends up expanding outward quickly and after just a few iterations you are indirectly processing information from everyone in the broader network. With the small-world feature of networks, it takes only a few iterations before information from one person has reached most others, at least in some diluted fashion.

This sort of learning in which each person repeatedly talks to their friends and simply keeps averaging their opinions is what is known as DeGroot learning, named after Morris DeGroot, a statistician.[12] The DeGroot model does not presume that people do arbitrarily complicated calculations, but just simple ones, like averaging numbers. Perhaps not surprisingly, when it comes to predicting how real people act, even in simple networks, the DeGroot model acts more like humans than a more omniscient and sophisticated model in which people adjust how they process information with time and in response to how they see others' opinions fluctuate, at least in some settings, as we shall see.

However, even DeGroot learning is still more complex than simple diffusion as it involves intensities and repeated conversations.

The centralities of people matter in how influential they are in steering other people's eventual beliefs. The friendship paradox (Chapter 2) is at work here. People who have more friends end up with their opinions being averaged by more people. In the network in Figure 7.1, the person with the black fill at the bottom of the picture is the most central by any centrality measure. Even though the average of all the starting estimates in this example would be accurate at 1,200 pounds, the fact that the dark-fill people with heavier guesses are more central than the white-fill people with lighter guesses makes the eventual consensus overshoot and overestimate the weight.

So, more central opinions have more of an impact. If you recall our discussion of centralities, you will realize that since people are talking again and again and again . . . what will really matter will not simply be someone's degree, but instead their eigenvector centrality. You are right. Being friends with other people who are well connected means that my opinion will spread to them, and then outward from them, and so having well-connected friends is as important as having many friends. Indeed, how much each person's initial opinion enters into the eventual overall opinion held by the society if the society keeps averaging repeatedly is *exactly* proportional to their eigenvector centrality.[13]

If the society keeps going through this process, it will eventually reach a consensus. The intuition behind this is seen in Figure 7.1 as the shades of the different nodes will eventually come to be the same. The person with the darkest shade will eventually lighten, and the person with the lightest will eventually darken. As long as someone is darker or lighter than their neighbors, they will be moved, and eventually the whole network will come to the same shade.[14]

What that consensus is depends on the starting estimates and the centralities of all the nodes. The eventual consensus belief is actually given by an amazingly simple formula: you just add up each person's initial estimate multiplied by their eigenvector centrality.[15]

There are some important biases in this process.

One is that there are "echoes": your own opinion gets reflected back at you. Your friends' opinions are partly based on your past opinions—so part of the "new" information you get from repeatedly talking to your friends are echoes of your beliefs. As your friends begin to confirm your beliefs, you can end up being overconfident. It is quite natural: you feel better about an opinion if other people agree with it. Even if you know a lot about the structure of the network, filtering out your echoes is hard.

A second, and even more extensive bias, is double counting. If you talk to both Lisa and Emilie, and they are each friends with Alex, then Alex's opinion is getting to you through two different channels. You end up double counting Alex's information. Hearing the same information from multiple sources makes it appear to be more reliable than if you hear it from one original source, even when they are just relaying the same information.[16]

Double counting and echoes are illustrated in Figure 7.2.

Biases driven by double counting information and not filtering out echoes are hard to avoid. You want to know if you should see the latest *Star Wars* movie. You read a review that says that the cinematography is spectacular. Both of your friends say the same thing. Is it really their own opinions? Were their opinions swayed by the reviews they read? Were they affected by the conversations they had since they saw it? When you eventually see it, and someone asks you how much you liked the movie, is your opinion now a bit higher because your friends loved the cinematography or because a critic

(a) The initial estimates, before any conversations.        (b) Two people are both influenced by Helen's
                                                                overestimate.

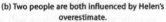

(c) Those two people both influence the last person, who    (d) Alex's effect is also partly echoed back to him.
    double counts the indirect information.

**Figure 7.2: DeGroot learning: Double counting and echoes.**

also suggested that it was great? Do you even know how much of
your impression of the movie is really entirely yours, and how much
of it was shaped by what you heard before and after? If you tell your
friend that you thought the dialogue was lousy, and they agree, how
much more confident in that opinion do you become? Are they echo-
ing your opinion or did they really think the dialogue was poor?

There are situations in which you will explicitly relay someone
else's opinion, and say that it is not from you. If someone asks me
where interest rates might be heading, it's not my expertise, but I
might quote the opinion of one of my colleagues who is an expert.
But even when I do, I often cannot explain where my expert friend's
information came from.

One way to directly see whether people are subject to double
counting, and failing to sort out echoes of their own influence, is to
do an experiment in which the information fed into the network is
fully controlled. If we know precisely what information people start
with, and what the network of interactions is, then in addition to
checking whether people fall prey to double counting and echoes, we
can also check whether DeGroot learning matches people's behavior.

Arun Chandrasekhar, Horacio Larreguy, and Juan Pablo Xandri,[17]
did exactly that. It turns out that DeGroot learning was amazingly
good at predicting how people's beliefs evolved.

The researchers put people in the experiment into networks. Whom each person could talk to—their friends—was assigned so that the experimenters knew what information came from whom. The networks were such that everyone could reach all of the other subjects in the experiment within a distance of at most four hops: there were friends, friends of friends, friends of friends of friends, up to a distance of four.

Chandrasekhar, Larreguy, and Xandri also controlled people's initial information and what they were trying to guess. There were two indistinguishable bags, each of which contained seven balls. One bag had five blue balls and two yellow balls—let's call that the blue bag—and the other bag had two blue balls and five yellow balls—which we'll call the yellow bag. The experimenters picked one of the two bags for the experiment, but the people in the experiment did not see which bag it was. Then each person privately got to see one ball randomly pulled out of the bag. If you happen to see a blue ball, then it is most likely that it is the blue bag (in fact, your best guess would be that there is a 5/7 chance that it is the blue bag). Instead, you might happen to see a yellow ball in which case you will think that there is only a 2/7 chance that it is the blue bag and a 5/7 chance that it is the yellow bag. After each person saw one ball randomly pulled out, in private, they each made an initial guess. After their first guess, they got to see how their friends in the network guessed.

After seeing their friends' guesses they got to guess again. Then they got to see their friends' second guesses. They got to guess again, and so forth. You might want to change your guess over time. If you initially guessed blue, but saw that all four of your experimental friends had guessed yellow, then presuming that your friends are paying attention to the game, you would infer that they all saw yellow balls. That would be four yellow balls to one blue ball, so you should switch your second guess to yellow. But each person in the network had different sets of friends, reaching out in the network. One of your friends might have initially guessed yellow but then later switched to blue. What should you infer from that? Also, it could be that some of them are changing because of you (echoes) or in reaction to a common friend (double counting). Sorting all of these guesses and changes in guesses becomes quite tricky after just a couple of rounds.

People knew what the overall network of friendships looked like, so they actually knew if some of their friends had a friend in common. So we can see if people acted sophisticatedly like a high-powered carefully programmed computer, filtering out echoes and avoiding double counting; or whether they acted more in line with DeGroot learning, which has people make guesses, without thinking too deeply, that just keep changing based on the most recent guesses of their neighbors.

There was a total of 665 participants in the experiment, put into networks of 7 each. So there were 95 separate networks, and many different guesses and situations for each person; overall there was an abundance of data on how people behaved. Thus the experiment had pretty good accuracy in terms of distinguishing how sophisticated people are.

Presuming that people are simply following DeGroot learning (with equal weight on each friend) correctly predicts 94 percent of the guesses; while presuming that people are fully sophisticated correctly predicts only 74 percent of the guesses. There are many guesses where things are easy to predict and so we should expect any model to match a good fraction of the guesses. For instance, on our first guess, we should each guess whatever color ball we saw. On our second guess, we should each guess what the majority of guesses were among our friends in the first round. The models don't really diverge until we get to later rounds where echoes can appear (my friends changed because of me), and double counting can appear (two of my friends are both changing due to seeing the behavior of some friend who is common to both of them), and so forth. That's where DeGroot learning starts to dominate: people do fall victim to double counting and echoes—and the simple DeGroot model is an amazingly accurate predictor of their behaviors and much more accurate than a fully rational and sophisticated model of behavior.

Other experiments also find that people fall prey to double counting and echoes, and have many other quirks.[18] Most important, this happens not just in small-stakes experiments, but also in hugely expensive decisions in our day-to-day lives.

Michael Bailey, Ruiqing Cao, Theresa Kuchler, and Johannes Stroebel tracked how people's decisions to buy houses were affected by interaction with their friends.[19] They tracked people's friends via Facebook and examined how the experiences of those friends influenced a person's decision. As an example, consider a person living in Los Angeles, let's call him Charlie, who is deciding on whether to buy a house or instead to rent. Charlie has a friend in Boston, let's call her Lucy, thousands of miles away. If Lucy's house has gone up in value, then it turns out that Charlie is more likely to buy a house, and will pay more, and buy a bigger house, than if Lucy's house has fallen in value. Charlie's decision is influenced by Lucy's experience with her house thousands of miles away. The study is careful to account for all kinds of possible confounds, such as people's characteristics and economic trends. The size of the effect is large on many dimensions: a 5 percent increase in a friend's house value over the past two years corresponded to a person being 3 percent more likely to buy a house, and to their buying a 2 percent larger house, and paying 3 percent more for it. Also, if friends' housing investments did poorly, then a person was more likely to sell their house and to take less money for it. In addition, the more variation there was in the outcomes of a typical friend of Charlie's experiences, the more cautious Charlie became. So, for instance, having two friends whose houses went up by 5 percent had more of an impact than having one whose house went up by 15 percent and another whose went down by 5 percent.

There are several interesting things to note. First, people's decisions are linked to their friends' information, even those who live far away. Second, this shows up sizably in important life decisions: buying houses. Third, it does not appear that people are properly processing the information they are getting from friends. Lucy's good luck with housing is from a market far away and based on what happened over the past two years. How predictive is this about what Charlie should be doing now? It turns out that people in the study did not adjust for where their friends were living—the Boston market is much less correlated with the Los Angeles market than the San Diego market is, but Charlie listened just as much to Lucy who lived in Boston as to Linus who lived in San Diego even though Linus's experience was much more informative.

Most of us are stuck with echoes, double counting, and paying attention to people whose information might not be relevant. Can we still manage to aggregate information well through our networks? Would we come close to assessing accurately the weight of the ox in a network in which each of us only talks to our friends?

It turns out, as Ben Golub and I found, that even a process as naive as DeGroot learning leads to very accurate eventual beliefs if a few critical conditions are met.

We have already seen some of the conditions: we need a diversity of views and they cannot be systematically biased, so it should be that if we were to average all of the opinions in the society the result would be accurate. If the society does not have the right information out there among its members to begin with, it has no chance. Beyond these conditions, the communication network has to be well "balanced." If all of us are friends with the same person, then his or her belief can dominate the overall outcome. The exact condition that ensures accurate learning is that the eigenvector centrality of each individual in the society has to be small relative to the sum of the centralities of the others. If we look at a network, what ensures this is a balance condition: roughly that the attention that people pay to any small group cannot greatly outweigh the attention that they place on others.

**Figure 7.3: Panel (a) an unbalanced network. Panel (b) a balanced network. The network in (a) is more efficient in having fewer links and having someone who has access to all the information in the network, but it can end up overreflecting the center person's views.**

We see an example of an unbalanced versus a balanced network in Figure 7.3. All of the unbalanced network passes through a single person. If that person places more weight on his or her own opinion than those of his or her friends (a natural tendency[20]), then that will be reflected back in the final beliefs. In contrast, the symmetry of the more balanced network means that nobody ends up driving the overall consensus that emerges.[21] The unbalanced network can save on communication costs, as it offers one-stop shopping for information: each peripheral person only needs to talk to one person to get information about what everyone is thinking, which is why one might see such starlike networks emerge.[22] But such unbalanced and centralized networks can end up with biased opinions.

Whom people pay attention to is something that has been extensively studied. For instance, in a series of studies in the 1940s and

1950s Paul Lazarsfeld examined how people formed their opinions. First was a study of 2,400 adults in Ohio during the months leading up to the 1940 presidential election. Lazarsfeld, together with two colleagues and a team of research assistants, interviewed a series of people repeatedly, and asked with whom they had talked, which media they were paying attention to, and what caused them to change their opinions whenever there was a switch.[23] Lazarsfeld also was involved with a later study of eight hundred women in Decatur, Illinois, asking them how they formed opinions on a variety of topics, including consumer products.[24] From these studies emerged a theory of "two-step communication" that was elaborated upon in a book by Elihu Katz together with Lazarsfeld. The two-step communication theory posits that there are people who operate as "opinion leaders"—mavens, as it were—who relay information gleaned from media to other people, "opinion followers." A similar theory was the centerpiece of Malcolm Gladwell's Law of the Few.[25]

Are there situations in which a society relies heavily on just one person's opinion? It is not hard to find examples. Some have suggested that Robert Parker, a famous wine critic, held immense influence within the wine industry. It is hard to dispute that Parker was "the" wine critic for several decades.

The wine industry is one in which information from such critics is vital. More than thirty billion bottles of wine are consumed each year, produced by tens of thousands of wineries many of which ship all over world. The quality of a wine depends on local weather and soil conditions, as well as how the vines are handled, when the grapes are picked, and how the wine is made. Even the same winemaker working from the same vineyards may have substantial variation in the quality of a wine from one year to the next. Putting all of this together with the high costs of many wines means that information from others about the quality and features that one can expect from any given wine is very valuable.

Robert Parker is a classic example of a self-made man. The son of a salesman, he founded his wine-rating newsletter in the late 1970s, at a time when he notes that there was a shortage of wine criticism—especially if you did not read French. For years he worked as a lawyer for a bank, and tasted wines and sold his newsletter on the side. But his was the right palate at the right time. Parker accepted no advertising and paid for the wines he tasted—he did not want to be swayed by any gifts or business ties to wine producers. He began rating wines on a 100-point scale, with a score above 90 generally indicating an excellent wine. Such a scale is now the standard in the industry.

Tasting wines takes talent, as one not only has to be able to discern what it tastes like now, but also what it will taste like when it reaches its maturity, often years after it is bottled. This involves detecting tannins, acid, sugar content, and various flavors, and understanding how they will evolve with time. Parker's influence began to really take off when he called the 1982 Bordeaux a great vintage, contrary to many others doing early tastings of the wines. It is now considered to have been one of the greatest vintages of any wine in history. As James Laube of the *Wine Spectator* put it, "along came 1982 in Bordeaux, with the most magnificent outpouring of great wines the world had ever tasted."[26] By the mid-1980s Parker had quit his day job and was rating wines full-time. Parker quickly emerged as the focal point in the industry. Max Lalondrelle, who buys wines for Berry Bros. & Rudd (a wine merchant in the U.K.), put it this way: "Nobody sells wine like Robert Parker. If he turns around and says 2012 is the worst vintage I've tasted, nobody will buy it, but if he says it's the best, everybody will."[27]

One effect of this is what many have called the "Parkerization" of wines. Winemakers sensing that Parker tends to like big wines— ripe, rich wines with strong flavors of the grapes, the oak barrels in which wines are often matured, the earth that nourished the vines, and a significant level of alcohol—have tended to tilt their wines in that direction. Elin McCoy's biography of Parker[28] quotes a Bordeaux merchant describing the impact for one winemaker's profits as saying "the difference between a score of 85 and 95 was 6 to 7 million Euro."

It can be wonderful to have a critic with such a discerning palate and memory for tastes and smells (and a nose insured for a million dollars) to recommend wines for the world. However, if you don't like the same sort of wines as Robert Parker, this can be bad news. Should we really be alarmed that one critic has enormous influence on wine sales? The wine industry has sales in the hundreds of billions of dollars a year, and having one critic's taste determine the success or failure of a wine introduces an enormous amount of uncertainty and risk into winemakers' lives.

Estimating the weight of an ox works better by having many estimates, even from pretty naive guessers, than just one estimate from someone who is fairly accurate but not perfect. The same is true of rating all sorts of things from wines to stocks.[29]

Evaluating wines and many other products is becoming easier. Even if none of my friends has tasted the latest vintage of some small-production Burgundy, there are people who have, and it is not hard for me to find their views online. Our networks for such information continue to grow and connect with people whom we have never met and will never hear from again, thus helping diversify the critical views that we draw from. This can help if the reviews really represent some independent experiences and new information.[30]

However, there are situations in which all the information traces back to one source. If that source turns out to have been unreliable, especially if it appeared reputable at the time, the results can be disastrous. This is especially true of matters requiring expertise.

On February 28, 1998, a leading British medical journal, *The Lancet*, published an article by Andrew Wakefield and twelve other researchers that found a link between autism and a vaccine for measles, mumps, and rubella—the MMR vaccine. Their explanation for what they thought might be going on was plausible: giving the three vaccines in combination caused intestinal problems and an immunity response, which ultimately led to problems with brain development in some children.

Given the worldwide rise in diagnoses of autism, and the number of parents whose children get the MMR vaccine each year, this was important news. The publication of the results in a reputable medical journal led to a firestorm of reporting on the subject. It was reported and re-reported, and became a widespread topic of conversation. As a first study on this explicit connection, it would be some time before the medical community had a closer look at the connection. Basically, the information originated entirely from one source in the network—the *Lancet* article. But people were hearing about it everywhere.

As it turned out, there were several serious problems with the study. First, it was based on just twelve children who were selected for the study. As autism can often appear around the same age as children get vaccines, it is natural to see an association between the two, and with a sample of twelve it is difficult to judge much of anything. Moreover, inferring any causation from such a study is clearly impossible. Although the article did not claim causation, it put forth a theory for causation that was misread by some as the conclusion of the study, and Wakefield suggested that the vaccines should be halted until further studies sorted out the correlation.

But there were even bigger problems with the study than the fact that it did not really establish any correlation or causation. Wakefield had received funding from a legal organization that was involved in suing vaccine manufacturers, which is considered a conflict of interest and of which *The Lancet* should have been informed. Most troubling was the report by Brian Deer, published in *The Sunday Times* on February 8, 2009, stating that the data were not accurate and did not match the actual hospital records, and that Wakefield had manipulated the data.[31] In 2010, the General Medical Council of the United Kingdom found Wakefield guilty of misconduct and struck him from the medical record (revoking his medical license) based on his conflict of interest and a finding that the study failed to act in the best interests of its vulnerable child patients.[32]

*The Lancet* fully retracted the article in 2010. By then a number of studies that involved larger data sets from around the world found no connection between the vaccinations and autism.[33] However, much damage had already been done. In 1998, when the study was published there were seventy-four reported cases of measles in the U.K.[34] The MMR vaccination rate in the U.K. soon dropped from over 90 percent to 80 percent. Estimates based on World Health Organization data are that, between 2000 and 2010, approximately five million children in Europe aged two to twelve went unvaccinated.[35] As we know from Chapter 3, lowering a vaccination rate even slightly can allow a disease to resurge. Predictably, within a few years measles cases went up by a factor of more than twenty. Routinely measles outbreaks were in the thousands by 2007 and stayed high until 2014.

The consequences of a dip in a society's vaccination rate are painful and can touch any of us. My close friend and coauthor, Toni Calvó-Armengol—some of whose contributions we have already seen in this book—died of the mumps in 2007. It was a year in which, with its slumping vaccination of children, Spain had over ten thousand reported cases of mumps. Such a sizable outbreak of a disease and its consequences, which could have been prevented by wider adoption of an effective and inexpensive (MMR) vaccine, are a tragedy.

Over time, the sheer volume of studies finding no relation between any vaccines and autism filtered into the conversations, pushing beliefs back and vaccination rates upward. Some of the increases in vaccination rates came from the reality of the diseases themselves. Once measles or mumps were resurging, people had to think harder about refusing to vaccinate their children. A discredited danger from the vaccine began to appear less worrisome than the real chances of a child getting the measles, mumps, or rubella. The higher the stakes, the more incentivized people are to actively seek and filter information, rather than let opinions come to them.

## The Changing News Landscape

*"Falsehood flies, and the Truth comes limping after it."*

—JONATHAN SWIFT[36]

*"Democracy Dies in Darkness"*

—THE WASHINGTON POST

As the vaccination example makes clear, having an informed population requires that quality information be disseminated widely.

The Internet has brought with it some interesting twists on how information is both produced and spread. Ironically, the ease with which information spreads can have a negative impact on how it is produced. There are two main effects of the ease with which people can package and repackage information.

If I wanted to, I could set up an official-sounding organization tomorrow and, say, call it "The Global Vaccination Information Center." I could build a Web site with an official-looking logo—making it look like it was scientific, or solid in some way—and I could start posting whatever information I wanted. I could build a series of such sites, and have them reference each other. I might also find other people with the same perspective who are writing about the same topics, and cite them, and they might start citing me. If when people look for "mmr vaccine side effects," they start hitting my fictitious site I am in business. It is relatively inexpensive and can reach people around the globe. I could push my view even if it happened to be terribly distorted.

The point here is not about whether there are any dangers associated with vaccines, but about how information is produced and spread. The more sources of information that pop up on any subject that are intertwined and pushing unreliable information, the harder it becomes to learn the truth. It might so happen that it all balances out, but even if that happens, learning is slowed down when there is a higher ratio of noise to content. Providing fake news can also be used as a political weapon, to incite distrust and hatred between groups, and to move groups to action.

The Pakistani minister of defense Khawaja Muhammad Asif once wrote: "Israeli def min threatens nuclear retaliation presuming pak role in Syria against Daesh. Israel forgets Pakistan is a Nuclear state too AH."[37]

What was that all about? The minister of defense of Pakistan was reminding Israel that Pakistan has nuclear weapons, and appears to have been saying this in reaction to thinking that the Israelis were threatening to use nuclear weapons against Pakistan. An Internet "news" Web site had published a story with the headline: "Israeli Defense Minister: If Pakistan sends ground troops to Syria on any pretext, we will destroy this country with a nuclear attack." It seems that the Pakistani minister of defense believed the story and thought that Israel was threatening that if Pakistan meddled in Syria, then Israel would launch a nuclear attack on Pakistan. The tweet in reaction leads one down a scary path of "tit for tat" (retaliation in kind), when there was no "tat." The Israeli defense minister never said what was quoted in the story, and in fact the story did not even have the Israeli defense minister's correct name.

The Pakistani defense minister is not the first to be fooled by what he thought was real news. Hundreds of thousands of people were famously frightened on October 30, 1938, when Orson Welles led a broadcast of an adaptation of *The War of the Worlds* on the radio. They were fooled by a series of simulated news bulletins. For almost half an hour, the radio station did not interrupt the bulletins to tell the audience that they were all fiction. People who tuned in after the introduction were in for a shocking news stream covering an invasion from Mars. The broadcast bounced from news of explosions on

Mars, to the landing of a cylinder in New Jersey, to interviews with military and government officials, updates on evacuations, and various other news flashes—all designed to sound like authentic news reports. A headline of *The New York Times* the next day read "Radio Listeners in Panic, Taking War Drama as Fact."[38] The broadcast snarled phone lines as many people called each other, the police, newspapers, and the radio station, trying to make sense of the broadcast. Early estimates of a million or more panicked individuals overstate the reaction, but there were likely hundreds of thousands of temporarily frightened people.[39]

When fake news is less absurd than an invasion by Martians, it becomes yet harder to tell fiction from fact.

Even major news services stumble from time to time. Would you believe a story with a headline "Blondes to Die Out in 200 Years" if it appeared on the BBC news site? (Perhaps it should have said "Dye Out.") The story[40] was based on a study by "German scientists" claiming that, due to blond hair being a recessive gene, blonds would become extinct within two hundred years. The story also predicted that the last blond would be born in Finland. No, it was not published on April Fool's. As it failed to provide any detailed reference to the "study" that it discussed, it was hard to trace its origins. The study was also re-reported by other major news sources, such as CBS, ABC, and CNN, and attributed to the World Health Organization (which denied it). Mention of the fictitious study seems to have first appeared in a German women's magazine called *Allegra*, which quoted a World Health Organization scientist who never existed.[41] *Allegra*'s story was later relayed on the German wire service, which may have been what led it to be picked up by the other news services. Stephen Colbert, later making fun of the fiasco, proposed that selective breeding be used to save the blonds.

So, how well does a typical person do at sorting fake from real news on the Internet? Let's set aside errors by major news providers and just examine telling clearer fact from fiction. Sam Wineburg and some of his colleagues from the Stanford History Education Group[42] tested whether students from middle school through college were able to infer the reliability of various information that they saw on the Internet. They were given tasks such as answering whether a posted picture of wilted daisies really offered proof of fallout from a nuclear disaster, discerning which stories on a site were news and which ones were advertisements, and discerning the reliability of articles on sites of professional medical organizations with varying motivations. The report summarizes the findings by saying, "When thousands of students respond to dozens of tasks there are endless variations. . . . However, at each level—middle school, high school, and college— these variations paled in comparison to a stunning and dismaying consistency. Overall, young people's ability to reason about the information on the Internet can be summed up in one word: bleak."

Inaccurate news that enters our networks slows or even precludes learning by biasing beliefs away from the truth. However, beyond the proliferation of fake news, and sites that appear to be something that they are not, the production of detailed and accurate news is also under fire from another angle.

Technological changes have not only decreased barriers to entry, they have also increased the speed of delivery and updating. The combination of technological changes has immense benefits. All sorts of information about a myriad of facts is at your fingertips—or just a voice command away. Do you need information about a medical condition, or a recipe, a problem with an app, the current weather in Beijing, or do you want to learn more about Cosimo de' Medici's life? Answers are available in abundance, and are easy to find. Most are remarkably reliable and many can be accessed for free. The variety, depth, and quality of information at our disposal is astounding. Moreover, news can be reported quickly. If you want information about a celebrity's death halfway across the globe just hours or even minutes ago, you can access it.

How does one reconcile this cornucopia of information with a concern about fake news and a claim that incentives for investigative reporting are under fire? Information about a recipe, how to get some app running, the weather, or Cosimo's life are not controversial. They have no real policy implications. If you want information about rehabilitating a knee after surgery most sites will have similar information, but if you start looking for information about birth control you will see variance. If you start looking for information about the conduct of a prominent politician, a government agency, or some private company, things can get even dicier. To get objective views on a variety of such topics, we rely on journalists to dig up information and to sift through, sometimes mountains of it, and to paint a fair picture of what is going on.

Newspapers have a long history of breaking important stories, and some remarkably timely. Clare Hollingworth had one of the most spectacular starts to a journalistic career imaginable. In 1938 and 1939 Hollingworth was doing volunteer work in Poland, helping Czech refugees escape from lands annexed by Nazi Germany under the Munich Agreement. In August 1939, on a trip to England, her passion for writing and knowledge of the region around the German-Polish border sufficiently impressed Arthur Wilson, editor of the *Daily Telegraph*, that he hired her directly. Her first week on the job had her flying back to Warsaw. From there she was assigned to head to the southern part of Poland, in Katowice, near the current border with the Czech Republic—and the lands that Germany had taken over in 1938. Germany had closed the border and Hollingworth saw that only diplomatic cars were permitted to cross. Clare borrowed a car, complete with Union Jack flags, from her friend the British consul, with whom she had worked closely in helping refugees escape. She set off across the border to investigate. On her trip back, she passed along a valley in which German troops and tanks were massing. One of the large camouflaging covers blew off in the wind, and Clare was there to see what lay beneath. Upon her quick return to Poland, she wired the story that was to become the first of her many scoops, and one of the most important stories of the time. The headline of the *Daily Telegraph* on August 29, 1939, read "1,000 tanks massed on Polish border. Ten divisions reported ready for swift stroke." She did not have to wait long for her second big story. Two days later, when she was back in Katowice, war erupted, and she was wakened by the explosions and gunfire. Hollingworth had many more scoops and adventures. She was the first to interview the Shah of Iran in 1941, she identified British agent Kim Philby as a Soviet spy in 1963, was an early predictor of a stalemate in the Vietnam War, and she opened the Beijing office of the *Telegraph* in 1973. It is said that even though she died at the ripe age of 105, she always kept a passport within reach in case a story called.[43]

As more people get their news via various social media and news aggregators, the ability for any news service to get revenue for its reporting has diminished. You might think that reputations for accuracy and careful reporting would still be rewarded. However, having people do the dangerous legwork that Clare Hollingworth did throughout her career no longer pays. A story from a reputable news service delivering high-quality reporting can be almost instantly quoted, repackaged, and delivered elsewhere. The news services that delivered careful reporting used to be rewarded by having a time advantage: if they got a story first, it was theirs throughout that news cycle. It would take a day, or at least a half day, before it could be re-reported. Having a reputation for always being on top of the major stories and breaking new stories built followers. Now that lead time is a matter of minutes. That speed erodes the incentives to be a news producer rather than a news repackager.[44]

This increasingly opens news to the Wild West of the Internet. Major social networking sites have advantages of scope. When people check up on what their friends are doing or saying, they can check news feeds that are tailored to their interests, which the site can infer. It's also one-stop shopping: catch up with your friends and get news at the same time. Will these sorts of media have incentives to invest in major reporting of their own? It's not easy to see why, as any story that they deliver can be repackaged—it would not be unique to their site. What will draw people to the sites is the social connections that they offer, and the news is packaged as the add-on, not as the main attractor. The emphasis becomes more on speed and matching topics to people, and there is less reward for providing careful and deliberate quality.

A report on the state of media by the FCC (Federal Communications Commission)[45] states: "An abundance of media *outlets* does not translate into an abundance of *reporting*. In many communities, there are now more outlets, but less local accountability reporting" (emphasis in the original).

The FCC oversees media in the U.S. and one of its charges is to promote competition and diversity in media. Its detailed report and concerns about the scarcity of investigative reporting seem well-justified. As an example, it cites situations in which reporting could have prevented disasters. One reference is to a mining disaster. As I found by looking into the details, the story is particularly disturbing:

Methane is commonly released in coal mining, as it is naturally stored in coal, and in fact coal mining is a major source of the methane released around the globe. When combined with the coal dust that is released during mining, it produces a highly explosive atmosphere. Ensuring the safety of a mine requires a series of precautions: proper ventilation to avoid a methane buildup, various spray systems that control coal dust, rock dusting (another method to control coal dust), and warning systems that monitor the atmosphere in the mine. At 3:27 p.m. on April 5, 2010, a thousand feet underground, in the Upper Big Branch Mine in West Virginia, a longwall shearer—an immense digging machine with enormous teeth—hit sandstone causing sparks.

The company used a code system to give employees advance warning of government inspections, actively concealed violations when inspections were imminent, and kept two sets of books: one for review by inspectors and the other only for company officials. Miners were told that raising safety concerns would jeopardize their jobs. As a result, there were many violations in the mine that day. Some of the critical spray systems were missing, and the company did not use adequate rock dust and so coal dust was at dangerously high levels. Supports were never installed in some parts of the mine, and their collapse limited ventilation, which led methane to accumulate. The methane detector on at least one piece of equipment had been deliberately rewired so that the equipment could operate illegally.[46] The methane that had built up in the area ignited and, combined with the extensive buildup of coal dust throughout the mine, caused a massive explosion that blew through miles of tunnels underground. Twenty-nine of the thirty-one miners who were working at the time were killed.

The Mine Safety and Health Administration's (MSHA) investigation found that the explosion was due to flagrant safety violations, and issued 369 citations and more than $10 million in penalties. Controlling the coal dust and methane that are common in such mines is not difficult, but it does cost money. It turns out that such citations were not new to the mine, despite its concealment of many of its safety violations. For instance, *The Washington Post* reported that the mine had been flagged for 1,342 safety violations in the past five years, and fifty times in the previous month. But the reporting was reactive: it was *after* the explosion. The government had been issuing citations and the mine kept ignoring or appealing them, over a period of years, until disaster inevitably struck.

There were three levels of failure. Clearly the company flagrantly and criminally ignored their responsibility to ensure their miners' safety. Second, the government system set in place to avoid such disasters had failed. In fact, there had been another explosion at the same mine due to a methane buildup and improper ventilation in 1997, and the MSHA knew of the significant methane buildup in the same area in 2004, but did not check whether the recommendations to fix the problem were ever implemented. Moreover, although there were numerous citations, the mine was allowed to operate for years despite its clear lack of safety.[47] This leads us to the third source of failure: a scarcity of reporting. We rely on investigative reporting to help unearth such failures *before* disasters occur. A free press ensures that public and private enterprises act in our collective best interest. When systems are corrupt or broken, it is hard to fix them if no light is shone on the problem until after the disaster.

In the year 2000, an estimated 56,400 people were employed in newsrooms at U.S. newspapers. By 2015, the number was down to 32,900.[48] Maybe that is not surprising, as newspapers have been under pressure from television and other media for some time. Moreover, many staff jobs have been automated, as now one does not need to talk to a person to place an advertisement, but can do it online. But the drop in numbers is not just due to the termination of nonreporting staff. For instance, television network news staffs have dropped by half since the 1980s, as have news magazine staffs since 1985. Even the fifty all-news local radio stations that existed in 1980 dropped to thirty by 2010 and reach only a third of the U.S.[49] A major source of news for many people remains local television news. However even there reporting is on the wane, and the amount of investigative reporting that local television delivers is limited. According to the FCC: "Topics like local education, health care, and government get minimal coverage. In a 2010 study of Los Angeles TV news by the Annenberg School of Communications, such topics took up just a little over one minute of the thirty minute broadcast."[50]

One way directly to measure how much investigative reporting is being done is to see how much information is being requested. In the United States, reporters (or citizens) can request information under the Freedom of Information Act. Between 2005 and 2010 the number of requests for information covered by the Freedom of Information Act dropped by almost 50 percent.[51]

In the world at large, newspaper circulation is dropping throughout North America and Europe, but actually growing in Asia. Economic growth, increased literacy rates, and low prices have increased circulation in China and India faster than circulation has dropped elsewhere, so that worldwide circulation is rising. However, revenues continue to drop, including digital revenues. Advertising revenues (including both print and digital) have plummeted, and revenues from digital subscriptions have grown much more slowly than news services had hoped. Overall, even though more than a third of people read newspapers online, revenue from all digital sources provides less than 8 percent of the industry's total.[52]

As the world becomes increasingly digital and mobile, where will the money come from to support investigative news production? It is not obvious how the Clare Hollingworths of the future will be paid. Although technology is amazing in the variety, volume, and speed of the information it provides, it is also becoming more challenging to earn revenue to pay for news that is hard to produce but easy to repackage. It tilts news production toward shorter, catchier, and more easily produced news, and away from the news that serves to police a democracy, which can be costly and time-consuming to collect.

As the FCC report suggests, it is not hard to see why David Simon, who reported for the *Baltimore Sun* for more than a decade before creating the HBO show *The Wire*, stated in a 2009 Senate hearing: "It is going to be one of the great times to be a corrupt politician."[53]

Our discussion has covered a number of sizable impediments to learning in human networks. But there is still the elephant in the room: homophily. No story of social learning is complete without it.

## Polarization: It's the Homophily Talking

People can live meters away from each other and belong to almost completely different social networks of generation, social class, ethnicity, religion, and, in many cultures, of gender.

Despite the breadth and speed of modern connectedness, we have not seen a decrease in the polarization in opinions and beliefs that people hold. In fact, there is evidence that political views are actually diverging in many countries.

One way to see this is to use textual analysis to track how the ways in which people express themselves are changing over time. One of my colleagues at Stanford, Matt Gentzkow, together with Jesse Shapiro from the University of Chicago, were pioneers in using such techniques to study biases and slants in media. They teamed up with Matt Taddy to analyze partisanship by analyzing people's speech patterns to quantify how partisanship has been changing over time.[54] They measure partisanship based on how easy it is to identify political party membership from the terminology a politician uses. When talking about immigration policy does a politician talk about "illegal aliens" or "undocumented workers," and when talking about a cut in taxes does he refer to it as "tax reform" or "'tax breaks for the wealthy"? It is not hard to guess which term is used more often by people from which political party.

Subtle as well as blatant differences in terminology reveal one's political leanings. It has not always been this way. Partisanship in U.S. politics was remarkably constant from the 1870s through the 1990s. However, after 1990 partisanship began a sudden and steep increase. For example, by counting how many times various snippets of words appear in one minute of a typical congressional speech from 1870 or 1990, a listener could be about 55 percent sure of the speaker's political party. So, one minute of terminology would provide a slight clue to a person's political leaning, but only 5 percent more than flipping a coin. However, by 2008 the terminology used by the different parties had diverged. By counting snippets of words used, and not the actual points made, one could be 82 percent sure of a speaker's political leanings. This is after only one minute of speech. After four more minutes of tracking words used in a 2008 speech, we can be more than 95 percent sure of a speaker's political party, while listening to speeches from 1990 or before leads only to a 65 percent certainty.[55] Although differences in the terminology are just one facet of polarization, they mean that our perception that politics is becoming more divisive is justified.

The increased partisanship in rhetoric is not just hot air: we also see more legislative gridlock.[56] Although parties have always disagreed, legislatures used to be places of compromise. Many of the bills that ultimately passed enjoyed bipartisan support. Votes now tend to be split more consistently along party lines. To measure this I pulled data on who voted on each side of a bill in Congress.[57]

In 2015 there were 339 votes in the Senate. The two senators who almost never voted the same way were Barbara Boxer, a Democrat from California, and Marco Rubio, a Republican from Florida. They voted the same way only 11 percent of the time. Basically, if you knew how Barbara Boxer was going to vote on a bill, you could be pretty sure that Marco Rubio's vote would be otherwise. The pair that agreed the most were Democrats Mazie Hirono from Hawaii and Jack Reed from Rhode Island, who voted the same way on more than 98 percent of the bills. In fact, the top thirty pairs in terms of voting the same way were all Democrats. Republicans were more split in their votes—a fracture that seems to be dominating the party's politics to an extent not seen since the Whigs imploded in the mid-nineteenth century. The important point regarding partisanship is how infrequently senators agreed with each other *across* parties, and how less frequent this is compared to 1990.

In Figure 7.4, two senators are linked if they voted the same way on at least half of the bills.[58] In 1990 that connects 82 percent of the senators to each other, while in 2015 it connects only 53 percent of the senators to each other—and very few of those connections remain across parties. An interesting feature of the networks is that some of the senators are pulled away from the main clumps of the parties. The positions of the nodes were not picked by me, but by an algorithm that I used to draw the network. This algorithm groups nodes closer together that are connected to each other and moves nodes away from each other if they are less connected.[59]

These patterns are not unique to U.S. politics. Nationalist parties have gained ground in countries like France and Austria. The Brexit vote uncovered a bitter divide in the U.K. A few years ago, Belgium spent a record 589 days without a government—with its Flemish-Walloon ethnic divide proving too wide. Spain spent much of 2016 with a zombie government, as its splintered parliament of regional interests was unable to put together a governing coalition.

How homophily affects the spread of information depends on the type of information in question. To understand this let's have a look at two types of information that we are constantly processing.

We help each other out by relaying pure bits of information that one does not need to interpret: "A bank will be offering microfinance in our village." "Our coach has been fired." "Our company is in financial distress and will be shutting our plant next year." "A new *Star Wars* movie is being filmed." These are things that you either know or don't—factoids or memes—all that is being transferred from one person to another is *awareness*. This sort of learning operates much

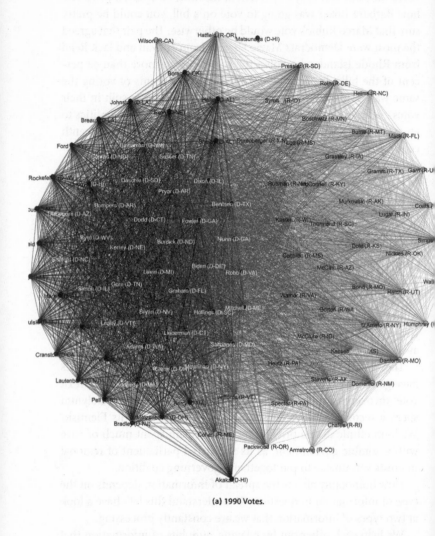

(a) 1990 Votes.

**Figures 7.4a–b: Votes in the U.S. Senate. Two senators are linked if they voted the same way on at least half of the votes. In 1990, 82 percent of pairs of senators were linked. In 2015, only 53 percent of pairs of senators were linked. The data are all votes in the Senate from GovTrack, which I scraped using Python code from Renzo Lucioni, as adapted by Peter Aldhous.[60]**

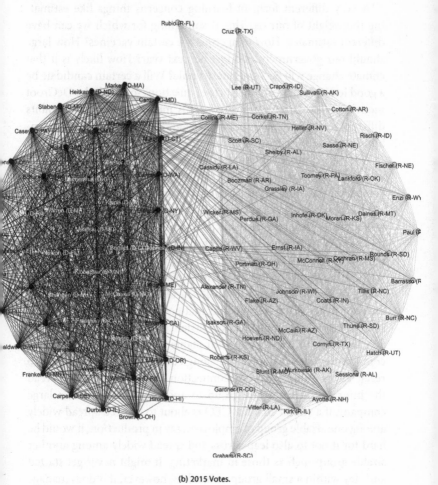

(b) 2015 Votes.

like the contagion processes we saw earlier. It is enough to talk to *one* person to become aware, and often the information is easy to convey.

A very different form of learning concerns things like estimating the weight of our ox. This is something for which we can have different estimates. How dangerous are certain vaccines? How large should our government's budget be next year? How likely is it that climate change will become catastrophic? Will a certain candidate be a good leader? These are the sorts of questions to which the DeGroot model applies. Learning of this sort is more complex and exhibits different learning dynamics. We are aggregating: assessing and combining information from multiple sources.

There are dramatic differences between how these two varieties of learning are affected by homophily—as Ben Golub and I found when comparing DeGroot learning to simpler forms of diffusion.

Homophily's effect on simple awareness of some factoid can be minimal. As long as a network is already dense enough to enable diffusion in its various parts, all that is needed is a small amount of interaction across groups to enable awareness of some fact to flow freely throughout the network. The explanation is quite simple, and in fact holds for contagion, diffusion, and learning settings in which the path structure of the networks matters most.[61] Suppose we have a society that is starkly divided, for instance a feudal society of nobles and peasants, or a company with plants in different locations, or a community that separates itself along ethnic lines. For awareness to reach across groups, only *one* connection needs to manage to spread the information from one group to another. If you are in a large company, if a rumor that the CEO is about to be fired spread widely among one sizable group of employees, say in production, it would be hard for it not to also leap across and spread widely among another sizable group such as those in marketing. It might never get started and stay within a small group of insiders; however, if it does manage to spread widely among any part of the company it will spread widely at more or less the same speed throughout the rest of the company too.[62] This echoes what we saw in Chapter 6: simple forms of contagion can easily breach schisms in social networks with nobles and monarchs dying alongside their subjects of many diseases. This is seen in Figure 7.5.

In contrast, homophily has a much more pronounced impact on more nuanced things like beliefs and opinions that take on a range of

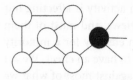

The starting condition: The five people who don't know the meme are separated by homophily from the people who do know the meme except for one connection.

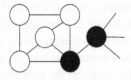

(b) The meme makes the leap and one of the five now knows it.

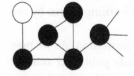

(c) The meme spreads to her three friends.

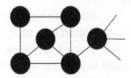

(d) The last of the five knows the meme.

**Figure 7.5: Spread of a meme with homophily: something that is either known or not spreads quickly, despite homophily.**

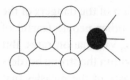

(a) The starting condition, the five white nodes are separated from the black nodes by homophily.

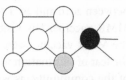

(b) The first person's beliefs are influenced by contact with the new belief, but only slightly, since they still interact with three others with her old opinion.

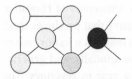

The influence spreads, but beliefs are only slightly moved given their starting condition and the homophily.

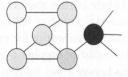

(d) The influence has reached the remaining node, but beliefs have not moved by much—it can take many more iterations to get beliefs to converge.

**Figure 7.6: DeGroot learning with homophily: Beliefs that can take on a spectrum of values shift slowly in the face of homophily—the relative weight on the new belief is low.**

values—such as the extent to which human activity is affecting our climate.[63] If I am aggregating my friends' views, and most of them start with similar views, then it can take an eternity for a minority view among a subset of my friends to begin to have an effect. We act like skeptics: the new information is going against most of what we are hearing and have been hearing. The process depends so much on repetition and inertia that a view only held by a small minority of our friends and acquaintances almost never has a chance to sizably shift our beliefs.

The difference between these two sorts of learning is seen by contrasting Figures 7.5 and 7.6.

The issue of vaccination provides a rich illustration. As we mentioned before, the reportedly fraudulent study published and later retracted led to a dramatic drop in vaccination rates. But that drop was not evenly spread. For instance, Minnesota is home to a tight-knit community of immigrants from Somalia. Although overall the state saw little reaction to the news about a link between MMR vaccines and autism, with vaccination rates remaining close to 90 percent, among the Somali community the vaccination rate dropped from over 90 percent to 42 percent by 2014. In fact, Andrew Wakefield, the lead author of the original study, visited the community several times between 2010 and 2011 to warn the population of the dangers of the MMR vaccine. His visits, and activism by other people fearful of the MMR vaccine, were effective in lowering vaccination rates. But the fear of vaccination was largely confined within the strong borders of the community, as were the dozens of cases of the measles that struck the Somali community in Minnesota in 2017.[64]

Another example of the impact of homophily on beliefs and behaviors is seen in the longline fishing industry, as studied by Michele Barnes and her colleagues from the University of Hawaii at Manoa.[65] Longline fishing refers to the industry that produces fresh tuna and swordfish for much of the world. Hawaii is a major hub in the industry, supplying over a thousand trips annually, and generating between $50 million and $100 million a year in revenues. The name "longline fishing" describes a technique. Fish are caught on long lines that have baited hooks attached at various intervals, often branching off via short lines. Hundreds and even thousands of hooks attach to a single line. For tuna and swordfish the lines are run near the surface. A challenge posed by the technique is that it not only catches tuna and swordfish, but also other species of fish, as well as sharks, turtles, and even birds. These are called bycatch.

Bycatch are undesirable to the fisherman for a variety of reasons. First, some of the species are endangered and so their bycatch is limited. Second, bycatch can be dangerous. As you can imagine, bringing in large tuna and swordfish is already dangerous, but detaching a shark from the line can be even more challenging. Third, each hook catching bycatch is one fewer hook for the desired catch, and time is wasted removing bycatch from the lines and repairing broken gear.

There are various ways in which bycatch can be avoided: changing baits, varying the depth of the lines, knowing how to spot high shark areas, knowing lunar and seasonal patterns of water conditions and bycatch locations, and sharing information with other fishers about current shark and other bycatch activity.[66] Thus, the fishers learn from each other.

Among Hawaiian longline fishers, the network of information sharing is very homophilistic. Michele and her coauthors interviewed nearly all of the fishers and asked with whom they commonly exchanged important information about fishing. The fishers break into three ethnic groups: Vietnamese Americans, European Americans, and Korean Americans (ordered from largest to smallest group). The Vietnamese American and Korean American fishers are typically first-generation immigrants and speak limited English, while the European Americans tend to be from the mainland U.S. Homophily along these ethnic divides is strong: 88 percent of the fishers' connections in the network are to their own ethnicity.

Barnes and her colleagues focused on shark bycatch, and found that Vietnamese American and Korean American fishers have similar rates of bycatch to each other, but rates that are significantly higher than that of the European Americans. Their estimate is that if all the fishers had the same (low) bycatch rate, ten thousand fewer sharks would be bycaught per year, corresponding to a 12 percent reduction of shark bycatch in the Hawaiian longline fishery.

Part of the differences in bycatch might have to do with cultural background. To see how much ethnic origin matters, Barnes and her colleagues dug deeper to look at how individual fishers behave. For instance, a few fishers of each ethnic background link to networks of some other group, such as a few European American, whose ties are predominately with Korean Americans. What predicts these individual fishers' bycatch is not their own ethnicity, but the network group with whom most of their ties lie. As further evidence that information is playing a role, Barnes and her colleagues, from interviews with the fishers, found that they share information within their networks about locations to avoid shark bycatch hotspots and discuss updated fishing technologies that improve efficiency.

As no direct information flows were observed in the study, we cannot be certain that homophily's impact on information is what caused differences in bycatch. Nonetheless, it does show us that different groups can maintain largely disjointed networks and different practices within the same industry, and in this case, in the same waters.

## The Challenges of Being Human

*"Only two things are infinite, the universe and human stupidity, and I'm not sure about the former."*

—ALBERT EINSTEIN

Whether Einstein said this or not, the quotation points out something important: our collective intelligence, impressive as it is, can get things wrong.

Our ability to grasp and communicate abstract thoughts has enabled us, as a species, to dominate our globe. Collectively humans have enormous knowledge, far beyond what any one of us could possibly master. But that same ability allows us to believe falsehoods. Our knowledge is not centralized and carefully aggregated by some omniscient being; rather it is decentralized and constantly changing. We process information from many sources. Very simple memes and ideas can flow easily, but we face significant challenges in processing and aggregating more complex information. Our networks are full of cycles and when we hear the same information from many channels, we are subject to double counting and can become convinced of information's veracity, even when it is just the same news coming at us via multiple paths in our networks. Coupled with homophily, and people's selective attention to sources of news, there is ample room for people to live in largely separate echo chambers and hold conflicting views across chambers.

Nonetheless, in spite of all of our human foibles, if: our networks are balanced enough, views in the population are centered around the truth, there are enough repeated conversations, and there are no strong preferences to manipulate or bias people's beliefs, then we can reach a consensus *and* get things right. The more charged or political a subject becomes, and the more people want beliefs to reach certain conclusions, the more selective people become in whom they talk to and what they repeat and the more homophily has an impact, resulting in polarization.

In addition to the complexities of how information is processed through our human networks, we face the challenge that technology affects not only the spreading of news but also the collection and discovery of news. Along with the extraordinary benefits of constantly making it easier to store, post, search for, copy, and broadcast information, technology changes incentives to produce true or false news. As the cost of producing and spreading fake but believable news gets close to zero, along with a vanishing reward to extensive digging for the truth, we need to become better at filtering information.

# 8 · THE INFLUENCE OF OUR FRIENDS AND OUR LOCAL NETWORK STRUCTURES

*"He that walketh with wise men shall be wise: but a companion of fools shall be destroyed."*

—PROVERBS 13:20

There are many reasons that our behaviors match those of our friends. Some behaviors exhibit strong complementarities: we prefer to use software and technologies that are compatible with those used by others. It is more entertaining to see movies and read books that we can discuss with others. We also imitate others because we infer that their choices are well-informed: a crowded restaurant signals good food. We follow the lead of role models—trusting in their discretion. As social beings, we also care about how others perceive us, and act accordingly—reacting to peer pressure.

These influences add another layer to how humans behave beyond simple contagion and opinion formation: people care deliberately to match the actions of others, resulting in additional principles concerning how network structure relates to diffusion and patterns of behavior. In this chapter we will explore herding and cascading behaviors, and gain more insight into why some products proliferate while others fail.

We will also dig more deeply into how and why local network structure matters. If we look at your friends, it turns out that there are important consequences to whether they are friends with each other. Beyond affecting diffusion of behaviors, these local network structures—such as whether two people have a friend in common—make a difference in whether those people can trust each other.

## Ants and Lemmings

> *"The human is indissolubly linked with imitation: a human being only becomes human at all by imitating other human beings."*
>
> —THEODOR ADORNO,
> *MINIMA MORALIA: REFLECTIONS FROM DAMAGED LIFE*

There are more than a million ants for each human on the planet,[1] and so it is likely that you live in an area populated by ants. In fact, unless you live in Greenland, Iceland, or Antarctica, you live near ants. Thus, you have undoubtedly noticed how quickly ants react to a food source. Drop a bit of ice cream on a sidewalk, and it won't take long before it is swarming with ants.

How do they do it? They don't have a central control system or brain—everything is done by individual ants. They interact through contact with others, in which they touch antennae and sense the chemical state of another ant, and via tiny releases of pheromones, which they can sense when they get near. One of my colleagues at Stanford, Deborah Gordon, has been studying ant behavior for decades, and she teamed up with a computer scientist at Stanford who is an expert on Internet protocols, Balaji Prabhakar, to see how these individual interactions end up driving the collective behavior. Together with a student, Katherine Dektar, they discovered what they nicknamed the "anternet."[2] The term anternet is appropriate because the system that the ants use for foraging has parallels to the ways in which the Internet is organized for the transfer of packets (small sequences of information). It is not surprising that nature discovered such a robust, simple, decentralized, and scalable algorithm long before humans did.

The species of ants studied is *Pogonomyrmex barbatus,* more commonly known as desert seed-eating red harvester ants, and they live in the deserts of New Mexico. Ants within a colony have different functions and the ones that find and bring food back to the colony are known as forager ants. The feedback system that regulates their activity is simple. Once a forager leaves the colony it keeps poking around until it finds food. Once it finds some food it hauls it back to the colony, often carrying things many times its own body weight. If forager ants are returning to the colony frequently, then foraging ants waiting near the entrance meet many returning ants. The ants that are waiting and ready to go react to that frequency, and the more times they bump into a returning forager, the more likely they are to go out themselves. Essentially each time they meet a returning ant they get stimulated, and eventually, when they are stimulated frequently enough per unit of time, they themselves go out to forage. When less food is available, the rate of returning ants diminishes and consequently the rate of ants leaving the colony to forage also diminishes.

As Gordon, Prabhakar, and Dektar recognized,[3] this system is akin to the TCP (Transmission Control Protocol) by which packets of information are sent over the Internet. When one sends a packet and receipt is acknowledged, one sends more packets. The faster the acknowledgments come in, the faster more packets are sent—basically the faster rate of acknowledgment signals that more bandwidth is available to send packets. If acknowledgments start slowing down, then the protocol slows transmission down.

The simplicity of this system makes it clear why humans, and many other animals, take cues from what is happening to their peers. If foragers are going out and quickly returning with a bounty, then it's a good time to forage: it makes sense that we are born with urges to imitate those who are successful around us. But beyond simply following others for when or where to forage, humans can also adapt the ways in which they cooperate and behave collectively. This ability to imitate and react to others is part of what enables our culture and norms to evolve without the much slower intercession of natural selection.[4]

But do people really behave as others have? Of course, we do it all the time. In fact, we do it even when the stakes are enormous. Lucas Coffman, Clayton Featherstone, and Judd Kessler examined whether Teach for America admittees were swayed by how others decided.[5] Teach for America selects high-achieving college graduates and pays them moderate salaries to teach in underperforming schools for two years. For example, a fresh graduate from Stanford might be assigned to teach math in a grade school in a poor rural area of Mississippi or in a lower-income neighborhood in Los Angeles. The program offers fresh graduates a chance to make a difference and learn more about the world around them, and helps schools fill holes with young, bright, and energetic teachers. A series of interviews helps Teach for America judge whether a candidate will be a good fit for the program, and also helps inform applicants about how big a commitment the program demands. After the interviews, if a candidate is accepted then he or she has to decide whether to commit or not. Coffman, Featherstone, and Kessler compared the rate of acceptance by admittees who were sent a standard letter congratulating them on making it through the qualifying interviews and asking them to join, to the rate of acceptance by admittees who were sent the same letter but with the addition of the following statement: "Last year, more than 84 percent of admitted applicants made the decision to join the corps, and I sincerely hope you join them." Eight percent of admittees who would not have joined without the statement joined with it.

This is only one bit of a mountain of evidence that people's decisions are swayed by those around them, but one with much higher stakes than usual—it's not about which toothpaste a person will buy, but how she or he will spend the next two years. Nonetheless, despite how obvious it seems to us that people imitate each other and coordinate their behaviors, it is not easy to be sure that they really do. Homophily causes headaches for a detective who wants to measure the extent to which our friends' behaviors influence our own. Do we act like our friends because they have somehow influenced us, or just because we are very similar to our friends and so we are naturally prone to act like them?[6] Did many people all go to the same movie because of learning via word of mouth and the desire not to be left out of the conversation, or was it because the movie happened to resonate among certain demographics, or because the same advertisements and positive reviews were seen by a particular group of people?

In order really to be sure that one person's decision *causes* another to make the same decision we need to have some control over the structure of the interaction such as in the Teach for America study. By randomly adding one sentence to the acceptance letter, we could be sure that the extra applicants who joined were reacting to the knowledge of how many others joined. But even when we cannot experiment, sometimes nature adds the randomness needed to discern what causes people to act the way they do. These are known as natural experiments.

For instance, weather patterns help us measure how many people will see a movie because others saw it. Suppose it happens to be blisteringly hot in Chicago, but pleasantly temperate in New York, on a summer day when a movie opens. More people than usual will go to the movies in Chicago, while fewer than usual will go to the movies in New York. The mere randomness of weather means that relatively more people in Chicago will have seen the movie than in New York in its opening weekend. What will happen the following weekend when temperatures return to normal? If people like to go to a movie because others are talking about it, then we should see relatively more people go to that movie in Chicago than New York. Indeed, for each extra hundred people who went to a movie on that first weekend, another fifty go the second weekend, and another thirty in the third weekend.[7]

Taking advantage of various forms of randomization, either by chance or by researchers, there are now many examples where we see people being influenced by the decisions and experiences of those around them, from whether Harvard Business School graduates choose to become entrepreneurs based on classmates' experiences,[8] to which apps people adopt,[9] to whether people enroll in a retirement plan,[10] to whether people exercise,[11] to which stocks people buy and sell.[12]

## Herding, Bank Runs, and Crowds Led Astray

> "By three methods we may learn wisdom: First, by reflection, which is noblest; second is by imitation, which is easiest; and third by experience, which is the bitterest."
>
> —CONFUCIUS[13]

The logic behind the Teach for America example, and others, is clear. Since a bunch of smart people before me joined, it must be a good thing to do. This is trusting in the wisdom of the crowd. But crowds can also go astray.

In the Teach for America case, the students learn not by deliberating with their friends and consulting a variety of sources, but by making inferences from what others have done. The danger is that such inferences can lead to herding behavior. How does a herd form, and what is so different about people making inferences from how a crowd behaves compared to talking to people one by one?[14]

Let's lower the stakes a bit; instead of Teach for America applicants let's consider people deciding between two places to eat: Alice's Restaurant and Monk's Cafe. Each person has some impression, for instance from looking at the menus or knowing the history of the chefs, about the qualities of the restaurants. Suppose that everyone would actually agree that Alice's Restaurant is the better choice if they had a chance to try both a few times. However, everyone's initial information is possibly erroneous, and people know that their information may be wrong and that their information is no better or worse than the information of others. There is real value in their initial information and if a large number of people were able to vote on which was the better restaurant, the majority would be correct with high probability.

But people don't get to all vote before choosing. Instead, they come one by one and choose which restaurant to go to. When they make their choice they can see what people before them have chosen (for instance, by looking in the windows).

The first person chooses a restaurant. By knowing that person's choice you can infer that person's information. If that person chose Monk's Cafe, then it is one piece of information in favor of Monk's so far. Now the second person chooses. Since the second person's information is as good as that of the first, the second person might as well choose the one they prefer. Suppose it is also Monk's Cafe. So now there are two pieces of information in favor of Monk's. But what if the first two people happen to have erroneous information and Alice's is the better choice? The third person to choose can see two people sitting in Monk's Cafe and nobody in Alice's Restaurant. You have probably been in this situation before: you are wary of going to a restaurant that is empty when one right next door has people in it. Here, seeing two people in Monk's Cafe leads the third person to infer that the first two people both thought that Monk's is the better place to eat. It is essentially two votes for Monk's, and so regardless of what the third person's information is, that person should know that the majority of information points to Monk's as being better. This leads the third person to go to Monk's, *regardless of her information*. This pattern cascades, as the fourth person (having read this book) realizes that nothing can be learned from the third person's information, but that the first two people both had information leading them to Monk's Cafe. This person should also choose Monk's regardless of his own information. This leads to herding—at Monk's Café even though Alice has the better restaurant.

This example makes an important point. If you are starting to think like an economist (and sorry if that scares you), you have an idea about what is going wrong here. There are externalities at work. The choice that any individual makes conveys information to others. People choose where to go for their own dining pleasure, but that ends up affecting what others can infer about the relative qualities. Just two people getting things wrong leads an entire herd astray.

The overall population would actually learn much more if the first twenty or thirty people had chosen based on their own information and had not seen what others had done before them. Then by seeing the first twenty people, one would then be able to infer twenty votes rather than just two. The fact that everyone past the first two did what was best for them—followed the herd and ignored their own information—meant that their information was lost to the rest of society.

Such "informational externalities" are rampant in our lives, for instance every time we face new opportunities or products. Which penguin wants to be the first to jump into the water when a leopard seal may be waiting? If we all stand around waiting, nobody will take the leap and we will learn nothing, or simply delay the learning process. Have you ever waited to learn from others' reviews or been scared away from a product that has no reviews? Sometimes that leads too few people to try a new product, which as a result never gets off the ground, or takes much longer to catch on.

In the adoption of new farming techniques this phenomenon has been quite costly. Adopting a new method can be risky; nobody wants to (literally) bet the farm on an uncertain technology. Andrew Foster and Mark Rosenzweig[15] showed how this sort of wait-and-see attitude substantially slowed the embrace of new high-yielding seed varieties in rural India. The poorest farmers were those who ultimately benefited the most from the new technology, but they suffered from having poor neighbors who were similarly reluctant to try the new technology.

There have been many disastrous products that consumers regretted trying—from Ford's Edsel to IBM's PCjr. But good or bad, a product has to be tried in order for one to learn about it. This is why manufacturers often hand out free samples of new products, and sometimes even pay people to adopt new technologies, especially particular consumer niches. Savvy marketers also take advantage of networks and offer referral incentives.[16] For a while, Tesla Motors paid an owner of a Model S sedan a $1,000 bonus if that owner referred a friend who bought a Model S. Dropbox's referral program helped it grow from 100,000 users in late 2008 to over four million by the spring of 2010. For some products, from movies to cars, there are reviewers who make their living telling others which to buy. And, of course, there are early adopters who will try some products no matter what and report on them, as well as people who have inside information and as a result are quick to get to the front of the queue.[17]

Externalities do not always play out badly, but they are double-edged.

If you see a long queue of people waiting to pull their money out of a bank, you might begin to worry about the safety of your deposits at that bank or even another bank. We could get it wrong: a few people mistakenly panic and start a queue to withdraw all of their money. We end up with a herd and its run on the bank without justification. Runs can happen completely on their own; however, they are usually sparked by some news. Consider the example in *It's a Wonderful Life*, in which Uncle Billy loses $8,000 of Bailey Building and Loan's money, causing concern and a line of customers to withdraw their savings, and a memorable turn of events.

As we discussed in Chapter 4, the news that sparks a run does not have to be consequential enough to ruin the bank, nor does it have to be true or related to the bank in question. It just has to sow uncertainty in the minds of the depositors and creditors about whether there will be a run on the bank. Fear of a run can cause a run. Even if you know that a bank has good investments and should be safe, if you anticipate that other people might panic, then pulling your money out becomes the prudent thing to do and a run becomes self-fulfilling. Even a healthy bank, when forced to liquidate its investments prematurely, may be able to give depositors only a fraction of their deposits back. Each of our decisions to pull our money forces other depositors to pull their money out—as none of us wants to be last. The externality here is that runs on a few banks can lead to the expectation of runs on banks we know to be perfectly healthy. Such widespread financial panics have become rarer in modern history—partly due to a trust in government insurance and an expectation that the government will intervene if needed. But there are still points at which even that trust breaks down and wider panic can hit good institutions along with bad—as in the suspension of banking in Greece in the summer of 2015.

## Game Theory and Complementarities

That some people pulling money from your bank incentivizes you to do the same is just one example of the many situations in which our actions are steered by what our friends and neighbors are doing.

Sushi Dai in Tokyo is one of the best sushi restaurants in the world, by many people's accounts. Moreover, its prices are remarkably affordable for the quality of the food. But if you want to eat there, what you don't pay in money you will pay with your time. Routinely, the line stretches around the block, regardless of the weather. We ate there after an almost three-hour wait, which included a sunrise, as Sushi Dai sits next to Tokyo's famous fish market, which is abuzz in the wee hours of the morning. Our experience was certainly not unique, as you can find all sorts of advice about when to go to minimize your wait. One guide explained: "We joined the Sushi Dai queue around 3:30 a.m. and were about 50th in the queue. With the first sushi served at 5 a.m. and 12 or 13 diners making it inside the restaurant every 45 minutes, this meant a 4.5 hour queue for us in advance of our 8 a.m. meal ticket!"[18] The populous line at Sushi Dai, however, is not simply a herd, as in our example of Monk's Cafe. There are other forces driving the queue.

The food was excellent, but to tell the truth, when it comes to sushi I cannot distinguish between best in the world and pretty good. Bad sushi is easy to spot, but above a certain quality level, I am not really able to judge.[19] So why did we wait hours in the cold and dark to eat sushi for breakfast, when there are many other places without any wait that we could have eaten where the food would have been indistinguishable to our taste? Because if you go to Tokyo and see the fish market, you are supposed to go to Sushi Dai. What does that mean? It's a shared experience. As the guide's Web site says: "We did it! We queued through the night for a Sushi Dai sushi breakfast in Tsukiji Fish Market and, having savoured every single piece, we can say it was totally worth the wait! A visit to Tokyo is not complete without a sushi breakfast and Sushi Dai is one of the best." Other tourists go there, and if you talk about the Tokyo fish market you will inevitably be asked "Did you go to Sushi Dai?" Now I can answer, "Yes, and it was an experience."

The sort of multiplier that underlies the long queues at Sushi Dai can become unbelievably large under the right circumstances.

In July of 2012 a South Korean musician named Psy released a video titled "Gangnam Style." Psy was a pop singer who was well-known in South Korea. He had spent a short time at Boston University studying business, eventually switching to Berklee College of Music in Boston, but returned home without completing his degrees. What he really wanted to do was make music. He struggled for a while before recording a string of songs and videos that would become hits in South Korea. The humor in his music and videos also managed to get him noticed in Japan, especially after a successful appearance at a concert broadcast on Japanese television in early 2012. However, outside of those arenas he was virtually unknown. So when the "Gangnam Style" video was released that summer, nobody would or could have predicted that it would become not only the most popular music video that year, but the first video of any kind ever to top *one billion* views. Two years later, it would pass two billion views. Sure, the video is catchy, features original dancing and creative filming, but so do many other videos. It had positive reviews, but nothing that would predict its runaway success.

The limits on our time and attention mean that not everything can go viral and be a topic of conversation. There is coordination involved, and again we find externalities: we view a video or movie because our friends are viewing it. The downside of such effects is that questionable things can go viral, while gems languish in obscurity. The best content may never get enough people in the right concentration to start snowballing. If you want to predict whether any given thing will have runaway success, you should just say no. You would be right more than 99 percent of the time. Most things never stand a chance. With the coordination involved, innate quality and eventual success are largely decoupled. Quality helps, but it is far from a guarantee. The "Gangnam Style" video was lucky enough to become one that people liked to play at parties, and then talked about, and shared to the point at which people had to see it to avoid being out of the loop.

That you care to coordinate your behavior with that of your friends leads to multiple outcomes that can emerge and be quite robust. In the parlance of a game theorist, these are "multiple equilibria." The mutual reinforcement effects can overpower our own individual tendencies. There are abundant examples in which we coordinate and get stuck on behaviors that are not optimal: we use computer keyboards that have letters in suboptimal locations because we need to be able to use keyboards in many locations and because it is cheaper to produce identical rather than custom keyboards; we speak languages that are needlessly complex and full of exceptions because we have to if we want to talk to those around us; countries end up with different customs for which side of the street they drive on, making it a danger for distracted or jet-lagged tourists to even cross the street. Better alternatives never take over as the feedback and strong incentives to coordinate mean that these behaviors can be nearly impossible to change, even after we recognize that doing so would be advantageous.[20]

These interactions are different from social learning—we care to do what our friends are doing, because they are doing it and not just because of the information that we learn from their actions. It would be pretty useless to be the only person in the world to know how to play chess. Most of our hobbies and entertainment involve some social aspect, from joining a cycling or book club, to chatting about the latest episode of a series. Given the social groupings around our choices, our actions also serve to signal things to others about our character, aspirations, and identities.

The wide variety of forces that push us to make choices similar to those of our friends have strong implications. Wanting to talk with our friends about common experiences, needing to conform, trying to keep up with others, or just trying to do things that are compatible with what others are doing, all compel coordination and this adds an interesting layer to the way that networks operate.

This sort of coordination also means that homophily, yet again, plays a role. When we coordinate our behaviors with our friends, something can diffuse and spread in one part of a network and not in another, much like what we saw in our discussion in Chapter 6 of decisions to drop out.

Corruption is actually a very social behavior—we are much more likely to ignore laws when others around us flout them as well—and a prime example of how we can see different norms across social divides.[21] Given strong levels of geographic homophily, the social aspects of corruption lead to very different levels of it across the world. This leads to some interesting discord when people travel, as they often take their own norms with them.

Ray Fisman and Ted Miguel found a fascinating example of this.[22] You have probably heard of "diplomatic immunity": diplomats can break some local laws without being prosecuted. As diplomats might not be fully aware of all local laws, this gives them some leeway; and it also prohibits a country from pressuring visiting diplomats with threats of arrest.

Most countries in the world have diplomats in New York, especially since the United Nations is there. Diplomatic immunity means that there is little that can be done if they park wherever they want to and then ignore the tickets. Between 1997 and 2002, diplomats from around the world accumulated over *one hundred and fifty thousand unpaid* parking tickets, with fines totaling more than $18 million. That's a lot of parking tickets. Most tickets were parking in illegal zones (no standing—loading zone, fire hydrant, etc.).

The difference in norms by country of origin is quite stark, with diplomats behaving very differently in concordance with norms in their countries of origin. If they experience corruption routinely in their home countries, then they are much more likely to ignore local laws and to get tickets. Some countries had *zero* violations, and you can likely guess which they are: Canada, Denmark, Japan, Norway, Sweden. These are all countries with low corruption. The countries that had *more than a hundred violations per diplomat* were ones much higher on the corruption index, like Egypt, Chad, Sudan, Bulgaria, and Mozambique.

Figure 8.1 shows the relationship.

"Corrupt" countries (those with positive corruption indices) averaged almost twenty-three unpaid parking tickets per diplomat while

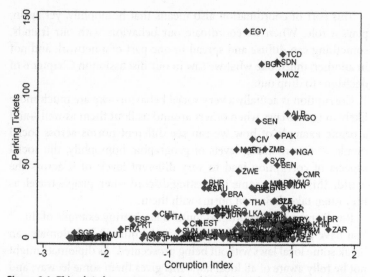

**Figure 8.1: This plots how many unpaid parking tickets there were by a country's diplomats in New York City between 1997 and 2002. The number is the number of unpaid parking tickets per diplomat. The corruption index is normalized to be 0 on average, and so a score of 1 means that a country is one standard deviation more corrupt than the average country, and a score of -2 means that a country is two standard deviations less corrupt than a standard country. Data from Fisman and Miguel (2007).**

noncorrupt countries averaged just over twelve unpaid parking tickets per diplomat.[23]

## Clustering and Complex Contagion and Diffusion

Diffusion of such complex behavior differs from the simpler diffusion of diseases and basic information, and even from social learning.

Before learning to play a game like bridge or mahjong, you might wait until several friends are playing before you make the leap. People feel safer asking for bribes when they see multiple others doing the same. Before joining a new social networking platform, you might wait until you have at least a few friends on it.

Diffusion in the face of social influence and coordination with others depends on more than the basic reproduction numbers we saw in Chapter 3; it also depends on what the local patterns of those connections are. A key network characteristic that captures relevant local patterns is called "clustering."

What fraction of your friends are friends with each other? This is called your "clustering coefficient."[24]

If Ernest and Henri are each friends with Gertrude, are Ernest and Henri also friends with each other? The frequency with which this is true for pairs of Gertrude's friends is Gertrude's clustering coefficient. The average of this across all the people in a network is the network's clustering coefficient.

The clustering of a network is pictured in Figure 8.2.

Clustering can depend on the type of relationship in question. For instance, in the data from our Indian villages (that we discussed in Chapters 2 and 5), the clustering coefficient is .22 for exchanging advice and .29 for exchanging favors (lending kerosene and rice). So there is more clustering in the second type of relationship.[25] It is natural that clustering will depend on the type of relationship in question. If you exchange advice with people from many different parts of your life, neighbors, family, co-workers, friends from school, and so forth, many of them might not know each other. If, in contrast, you only borrow money from your closest friends and family, then they are more likely to know each other.

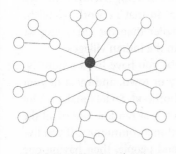 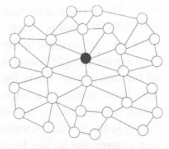

(a) A "tree" network with no clustering.          (b) A network with substantial clustering.

**Figure 8.2: Clustering: Two networks with very different clustering. In both cases, the solid node has seven friends. In the tree network in panel (a) none of the solid node's friends are friends with each other, and so there is no clustering. In the network in panel (b) one third of that person's pairs of friends are friends with each other (seven pairs of the solid node's friends are friends with each other out of that node's twenty-one pairs of friends), and so her or his clustering is 1/3.**

These clustering coefficients of .22 and .29 are not atypical; substantial clustering abounds in networks of social relationships, from the Florentine business and marriage network, to the adolescent friendships we have examined.[26]

To see what is special about these numbers, let's put such typical clustering in perspective.

Let's define a friend of yours to be someone who has done you a favor: lent you something, helped you with a task, or given you important advice. You might have about a hundred friends according to this definition. Someone living in a small, remote village with strong religious and gender segregation might have only a dozen such friends, and someone like Gertrude Stein who traveled the world and held popular salons might have several hundred. There is variation, but let's suppose that the average person has a hundred friends. As will soon become clear, you could make it thousands and that would not change the point.

Out of the world population, which is heading toward eight billion people, on average a person would be friends with one out of eighty million. Suppose that the world's friendship network had this frequency of friendships, but with every pair of people having an independent and equal likelihood of being connected. Then any two people would have a one in eighty million chance of being friends with each other. In such a network, any two of your friends would have a one in eighty million chance of being friends with each other. The average clustering coefficient in that network would be 1/80,000,000. That is clearly much smaller than the clustering we often see, such as those in the .2 range in the Indian villages.

Geography might matter a lot. Maybe you have lived a nontrivial fraction of your life in the same city or town, and most of your friends live in that same town or neighborhood. So let's make it a lot easier. Let's not form a network completely at random out of all the people in the world, but instead just within a community. If you live in a community of, say, twenty thousand people, then having one hundred local friends would mean that you are friends with about one in two hundred of the community's population. If we formed such a friendship network completely at random within a community of twenty thousand people, with a probability of any two people being connected at one in two hundred, then the chance of two of your friends being friends with each other would still be one in two hundred. Average clustering of 1/200 is still relatively close to zero and way off from clustering coefficients of one fifth or more.

Homophily also contributes, at least a small amount, to clustering. If we start making most of your friendships with other people who are similar to you, say in terms of age, gender, education level, religion, and so forth, then we can get things up to about 1/20, but it still falls short. Arun Chandrasekhar and I did this with our Indian village data. Even if friendships were predicted based on detailed homophily and allowed to depend on all sorts of demographics, we could only get clustering up to about .05, still well short of the .22 to .29 clustering in the villagers' networks.[27]

Beyond all the segregation patterns resulting from homophily, friendships still *cluster*. It is not hard to explain why we see so much clustering. We often interact in small groups. Students take classes together, play sports on teams together, and hang out together in small posses. People often work together in small teams or in shifts. Also, you meet people via your current friends—you know them precisely because they are a friend of a friend, and that also increases the chance that you become friends.[28]

Now that you know what clustering is, let us see how it matters. In fact, you have already seen that it can lead to double counting in social learning. Highly clustered networks have lots of cycles in them—lots of chances for the same information to reach me via multiple paths, and corresponding opportunities for me to double count, and to hear echoes of my own views. But beyond distortions in social learning, the fact that humans cluster their relationships also has a big impact on the diffusion of behaviors in which someone needs to be influenced by multiple contacts before being willing to act.

To understand how this works, suppose that two friends invent a three-person game. Other people learn to play the game if at least two of their friends do. Let us contrast diffusion of this game in the different networks we saw in Figure 8.2: a tree with no clustering, and the same network in which clustering has been added. The two inventors are the dark nodes in Figure 8.3. In the tree network, nobody is friends with both of them, and so the game never diffuses.

If instead, the network has clustering, as in Figure 8.4, then the game diffuses. It diffuses more slowly than a simple contagion process in which just one friend is enough to spur infection, but it still diffuses.

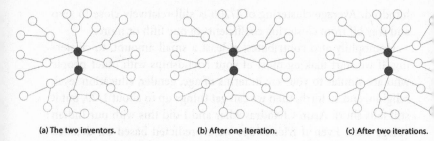

(a) The two inventors.

(b) After one iteration.

(c) After two iterations.

**Figure 8.3: Diffusion without clustering: If people need to be influenced by multiple different friends before acting, then without any clustering there is no diffusion. The game never diffuses.**

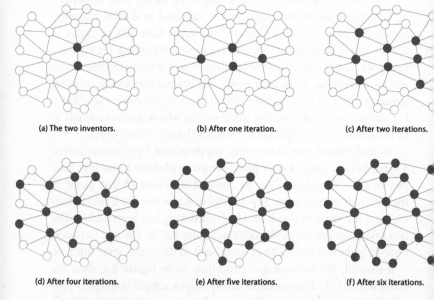

(a) The two inventors.

(b) After one iteration.

(c) After two iterations.

(d) After four iterations.

(e) After five iterations.

(f) After six iterations.

**Figure 8.4: Diffusion with clustering: Clustering enables diffusion when people need to be influenced by multiple (here two) different friends before acting.**

The example of a three-person game is a stark one, but there are many examples in which local clustering and network structure matter,[29] from the diffusion of language[30] to the adoption of new technologies.[31]

Clustering is not the only important feature of our local networks.

## *Trust and Support*

John Nash Jr. was an amazing mathematician. By the age of twenty-three he had published two short articles that transformed game theory,[32] laying a foundation for much of its eventual use in the social sciences. By the age of twenty-four he had proven an important embedding theorem in algebraic geometry that bears his name. He would go on to make creative and original contributions to the study of partial differential equations, implicit functions, and singularities.

His mind seemed to be wired differently, and unfortunately, in his thirties his extraordinary career was cut short by schizophrenia. Much of his adult life was spent in Princeton. Recovering from schizophrenia is unpredictable, and less likely the more time one has spent exhibiting the sorts of episodes that John Nash did. But people can and do emerge from the disease. The accepting environment of a university, together with his support network of friends, enabled Nash to survive for the decades in which he lived in a different world. I first knew him as a homeless-looking man who roamed the library and corridors of Fine Hall (the math building at Princeton), often with large stacks of papers. A friend informed me that the "Phantom of Fine Hall" was the same man I was learning about in my math and economics courses. John Nash did emerge from schizophrenia in his sixties, and went on to be recognized with an Abel Prize in mathematics and a Nobel Prize in economics. His incredible life was the story of Sylvia Nasar's book *A Beautiful Mind,* and Nash was portrayed by Russell Crowe in the Academy Award–winning film of the same title.

A small but important piece of trivia from Nash's life is relevant for our discussion. Princeton's math department was especially strong in the areas in which Nash went on to make his contributions, arguably the best in the world. As they admitted only a few grad students per year from a pool of many talented applicants, it was not always easy to decide whom to admit. A key to Nash's admission was a three-sentence letter of recommendation from Richard Duffin. The full body of the letter was:

"This is to recommend Mr. John F. Nash, Jr. who has applied for entrance to the graduate college at Princeton. Mr. Nash is nineteen years old and is graduating from Carnegie Tech in June. He is a mathematical genius."

The strength of the letter came from Duffin's reputation, which would have been known to the faculty at Princeton. Duffin was trained in physics, and working in mathematics and making important contributions to Fourier series and number theory, network theory, and geometric programming. We all might know people whom we think of as "mathematical geniuses" because they are very good with numbers, or incredibly knowledgeable on the subject. But these are not people who are going to make paradigm-shifting contributions to multiple areas of mathematics. When Duffin said that Nash was a genius, he really knew what he was talking about. The letter was so short since nothing else needed to be said. Duffin knew that Nash was a mathematical genius and that was all the admissions committee needed to know.

The relationship between Nash and Princeton was, at least in part, established and supported by Duffin's reputation and relationship with faculty at Princeton. This sort of connection is common in networks.

Let us say that a relationship between two people or nodes in a network is "supported" if they have a friend in common.[33] Nash's relationship with Princeton was supported by Duffin. This kind of support is vital to much of our interaction, and we already saw some of its impact in our discussion of job contact networks—but it matters well beyond our jobs and admissions.

Humans rely on each other in many ways to survive. Our children are helpless on their own for many years, and they rely not only on parents but also on others for their growth and safety. At various points in our lives we are in need, and have to borrow from others.[34] Early in our history, humans had to share and cooperate while hunting. Hunting large animals was sporadically successful. Since we are not like lions who can go for weeks without food, while some members of a community hunted for large but irregular supplies of protein, others had to provide more stable calories by gathering fruits, nuts, and other foods.[35]

Such informal relationships, in which we provide favors and lend to one another, rely on a whole series of incentives.[36] When a friend comes to ask you for help or a favor, you don't ask him or her for a cash payment or to sign a contract saying that they will pay you back. When we ask a friend to help us move, to watch our children, to give us advice, to help us financially, or for aid in an emergency, offering to pay for the friend's favor would be offensive and call into question the very nature of the friendship. Most of our interactions are not based on formal contracts, but on trust. That trust comes from a history of repeated interactions and patterns of reciprocity with the same individuals, as well as having friends and a community in common.

Gossip and rumors are things that we often think of pejoratively: people talking behind each other's backs. We start from an early age, tattling on a sibling, and hone the skill by the time we hit our teens when we relish in knowing who is doing what with whom. Although aspects of gossip seem intended purely for the pleasure of conversation, gossip also serves a higher purpose. When a person misbehaves, information about him or her can spread through a network and damage the person's reputation. The speed with which gossip can travel means that we know that treating a friend poorly can quickly be known to many others in our networks, especially when our relationships are supported, so that friends in common are likely quickly to be in the know.[37]

Concern about what other people think of us is a powerful motivator. People act differently when their actions are more widely observed—it is most natural. This gives us even more reason to expect that a relationship will function better if it is supported—if the two people have friends in common.

I worked with Tomas Rodriguez Barraquer and Xu Tan to study the role of such supporting relationships in enabling people to exchange favors.[38] Informal risk sharing is a good example to consider since it has a large impact on well-being, as mentioned in previous chapters, and because it relies on reciprocation and social pressure. When someone asks a friend for a large loan, the incentives of the friend to provide the loan and of the borrower to repay it depend on whether they have friends in common. The friends in common, as witnesses to the relationship, can cut their ties to a person who fails to help others when asked, as well as with someone who fails to repay their debts. The incentives for two people to borrow and lend from each other are thus enhanced if they have friends in common whom they risk losing by behaving badly compared to if their relationship is isolated with no friends in common.

This suggests that we should tend to see important informal relationships being supported by friends in common. We tested this in our Indian villages, and found that 93 percent of the borrowing and lending relationships are supported by a friend in common, which is 63 percent higher than the rate at which any two households who don't borrow and lend with each other are supported. It is extremely rare for two people to borrow and lend with each other without having a friend in common.[39]

Emily Breza and Arun Chandrasekhar thought, why not put this to work for people's own good? One challenge in getting people out of poverty is getting them to save money.[40] Setting money aside on a regular basis is not easy for most people and is especially difficult for the very poor. Beyond having a low and fluctuating income, there are always temptations and current needs. Moreover, saving small amounts of money can seem futile, even though it can be beneficial down the road. So, Breza and Chandrasekhar thought, what would happen if instead of just opening accounts for people and telling them about the future benefits of savings, people were also assigned a monitor? The monitor was someone from their community, who would do nothing except observe. Your monitor would just know if you met your savings goals.

Having a monitor turns out to make a big difference: it increased savings by 35 percent. People saved more by working more and by lowering their expenditures. Interestingly, it also made a difference who the monitor was. Monitors who were closer in the network to people they observed, and monitors who were more central, led to the highest savings levels. Beyond the average 35 percent increase, increasing the monitor's centrality by a standard deviation increased savings by another 14 percent, and having the monitor be one step closer in the network to the saver increased savings by another 16 percent.[41] Having people be aware of your actions can change those actions substantially, and even more so when that person is someone you know and is central in the network.

The role of supporting relationships is also key in business. A typical example is described by Brian Uzzi, who got to know the New York garment industry inside and out. He notes[42] how the CEO of a company that cuts fabric describes starting to work with a manufacturer named Diana. The CEO reported that a close business friend of both his and Diana's, named Norman, asked him "to help Diana out" by cutting her fabric at a low price in a hurry. The CEO reported: "What was my relationship with Diana? Really nothing. I didn't know if she had 10 dollars or 10 million dollars. . . . So why did I help her out? Because Norman asked, 'Help her out.' So when the account started, I gave it a hand. I cut the garment for 40 cents rather than what it was worth, 80 cents."

As Brian Uzzi describes it, "Norman's referral was the basis for the CEO's trust in Diana, even though she did not sign contracts, offer collateral, or guarantee return business." And, as a manager in Diana's company said, "There was no talk of 'one hand washes the other.' . . . It's understood here."

The two companies that both had strong relationships with a third party were willing to work with each other without any guarantees other than their friend in common, even though it was costly to do so.

## Guanxi

In China, such social capital, referred to as guanxi, is a vital part of business. Personal bonding and a host of past relationships and direct and indirect connections between businesspersons can be more important than a written contract. Such relationships take time to build, and so require patience and a long-term perspective. Disputes and misunderstandings are often worked out via the network rather than via litigation.[43]

Although guanxi can help people trust each other and work together, it also comes with a dark side with which we are all familiar: favoritism. We may prefer to offer our friend a job rather than a better-qualified stranger.

This sort of favoritism associated with guanxi was measured by Fisman, Shi, Wang, and Xu in the context of Chinese scientists.[44] As China has greatly increased its funding of scientific research, much of the money has flowed to scientists who are fellows in the Chinese Academies of Science and Engineering. Fellowship in the Academies is decided by a committee of current fellows. Fisman, Shi, Wang, and Xu measured a person's guanxi with existing members of the Academies by examining their hometowns. Having a common home-town with someone is an important form of guanxi called "laoxiang guanxi." The authors find that having laoxiang guanxi with a current member of a key committee deciding on who becomes a member of the Academies leads to a 39 percent increase in the chance of becom-ing a fellow (controlling for all sorts of other potential confounding effects). The fellows who are elected and have such laoxiang guanxi with a member of the selection committee are half as likely to have a high-impact scientific publication as those who got in without a hometown connection.

This ends up having consequences for funding: Fisman, Shi, Wang, and Xu estimate that Academies fellowship is associated with an increase in research funding of $9.5 million per year for the fel-lows' institutions.[45]

Guanxi not only distorts the outcomes, but also means that people end up spending a lot of their time building their networks rather than working. One cannot change where one is born, but there are other forms of guanxi and connections that one can control. When discussing the behavior of scientists, Yigong Shi and Yi Rao, state: "A significant proportion of researchers in China spend too much time on building connections and not enough time attending seminars, discussing science, doing research, or training students.... Some become part of the problem: They use connections to judge grant applicants and undervalue scientific merit."[46]

China is, of course, not the only place in which favoritism distorts productivity, and there are many studies on the subject in a wide variety of settings. The prevalence of hometown guanxi just makes favoritism easy to see since hometown connections are out in the open.

The full impact of the favoritism that comes with supported ties, guanxi, and other forms of social capital is nuanced because connections help foster trust, which can help in production, while also causing distortions in who gets what position and how much time they spend socializing.[47]

## Embeddedness

The fact that support and our local network structures matter so much also means that by analyzing a network one can predict identities. As an illuminating example, consider the following challenge. Just by looking at the structure of a network and without any identifying information, is it possible to identify which linked pairs are spouses or romantic partners?

Lars Backstrom and Jon Kleinberg took up this challenge using data from Facebook.[48] Here was the precise challenge. Pick someone, let's call him Sam, who is involved in a romantic relationship—either married or seeing someone regularly. Now, I show you a picture of Sam's local network—so his friends and which of them are linked to each other—and nothing else. You just see a map of nodes and links. From that, you have to guess which one of the nodes is Sam's romantic partner. If Sam has a hundred friends, and you guess randomly, you would end up with a 1/100 chance of picking correctly.

As a start, since having friends in common strengthens a relationship, providing support and enhancing trust, it makes sense that having many friends in common would be a good predictor of who is your spouse or romantic partner. The number of friends in common that a pair of people have is sometimes referred to as their relationship's "embeddedness."[49]

So you could look to see who among Sam's friends has the most friends in common with him. It turns out that if you do this with the Facebook data, you would be right 24.7 percent of the time, so about 1/4 of the time. That's impressive—much higher than just picking at random.

Going a step further does even better, as follows. Suppose that Sam is romantically involved with Marya. Their friends in common are a list of people: Nathan, Kelly, Luke, etc. What fraction of these people are friends with each other? Are Nathan and Kelly friends? Are Nathan and Luke friends? and so forth.

What might we learn from this? Suppose Sam and Marya had only met recently and knew each other through a skiing club to which they both belong. That is where they met, but outside of that their networks were quite separate. Then their friends-in-common would all be members of the ski club, and would be friends with each other. Suppose instead that Sam and Marya had known each other for years. Then only a fraction of their friends-in-common would be from the ski club, and others might be from each of their families, their jobs, their neighbors, their childhood friends, and so forth. Each of them has friends from a variety of sources, and over time they would get to know many of those friends. Since those friends-in-common between Sam and Marya then come from so many distinct places, many of them would be unlikely to know each other: Sam's friends among his co-workers would probably not know Marya's friends from her childhood.

Define the "dispersion" of the relationship between Sam and Marya to be the sum of all the distances in the network (ignoring Sam and Marya) between their friends-in-common. Having more friends-in-common adds to dispersion, as does having those friends not be friends with each other.

If you pick Sam's friend with whom he has the highest dispersion, then *60 percent* of the time you will have identified his romantic partner! This is more than twice as good as just picking the person with whom he has the most friends in common. By looking at the network with no other information, one has a better than one-in-two chance at identifying a person's romantic partner out of hundreds of people.

Interestingly, in cases where the network did not correctly identify the romantic relationship, it was often the case that the romantic relationship was doomed. In those cases in which a person's romantic partner was not his or her friend with the highest dispersion, then that romantic relationship had a 50 percent higher chance of breaking up in the next two months than if dispersion correctly identified the romantic partner.[50]

There is a strong logic as to why this works. High dispersion tends to indicate someone with whom we have spent a lot of time in a variety of different contexts. Our romantic partners are often the people with whom we spend the most time, especially when it comes to interacting in a variety of different settings. As a by-product a high level of dispersion also makes the breakup of any long-term relationship more difficult and acrimonious—as two people's networks have become increasingly intertwined over time.

## To Understand Us Is to Know Our Friendships

*"The people you surround yourself with influence your behaviors, so choose friends who have healthy habits."*

—DAN BUETTNER, CYCLIST AND EXPLORER

Our tendencies and desires to match our behaviors with our friends and acquaintances have many consequences. Our herds can help us find beneficial behaviors without having to spend too much time or energy exploring the options ourselves, and our shared experiences can lead to delightful discussions and interactions. But our herds and incentives to conform can also lead us astray as they cascade through a network, from the minor headaches of getting stuck with an inferior technology to the disastrous consequences of pogroms.

Once again, accounting for network structure helps us to better understand our behaviors and the spread of those behaviors. The diffusion of a behavior that is adopted only after multiple interactions depends on the amount of local reinforcement (e.g., triangles) in our networks. How one person treats another involves looking beyond that relationship to the network in which they are embedded. Are they on the same or different sides of a homophilistic divide? Do they share many friends in common who support their relationship and help foster their trust in each other, and provide incentives for them to exchange favors and to favor each other?

# 9 · GLOBALIZATION:
## OUR CHANGING NETWORKS

Dave Baun cycled the rural roads that cut through the hills not far from his home in Nazareth, Pennsylvania. Halfway around the world, Liisa Grace rode the many beautiful hills outside of Adelaide, Australia, known for their wines and the occasional kangaroo. It is hard to find two places as far from each other, so it was unlikely that their rides would ever intersect. Nonetheless, Liisa and Dave managed to meet.

Strava is a fitness app that allows a cyclist to track each ride, storing detailed data including average speed, distance, elevation, wattage, and calories burned. If cycling is your thing, such an app is a must-have. It helps you see where others are riding and makes it easy for you to find the best routes no matter where you are. It also allows you to see how others have performed on the same or similar routes. Strava helps match you to others with similar cycling habits so that you can compare fitness and track your progress, and send each other encouragement.

In 2014 Liisa sent Dave a request to follow him on Strava. He agreed. After exchanging kudos and pleasantries on Strava, they eventually connected via social media and deepened their friendship through long-distance conversations. Trips to Australia followed, and by 2016 they were married. Dave had proposed by riding a carefully planned route that spelled "Marry Me Liisa" when viewed on the Strava map.[1]

Platforms such as Strava are tremendous inventions that can bring knowledge and joy to our lives. Elderly who would be otherwise isolated can stay connected with friends and family via social media.[2]

Increased connectivity has also had widespread and unprecedented economic impact.

If you are a fisherman in a small village in rural China, India, or Africa, how do you choose where to sell your fish? It used to be that you had to just guess and go to a market and take whatever price you could get. Because markets were far apart and held roughly at the same time, it was not feasible to drive from market to market looking for a good price. But now you can call friends and acquaintances to find out what prices you might expect where. If there are too many sellers at one market and too few at another, then this helps steer a seller to the market that needs more fish. Does it make a difference? Yes, connectivity is an enormous equalizer. Rob Jensen looked at markets in southern India before and after cell phones became available.[3] Before cell service, prices for sardines would vary dramatically; for instance, the price in one market was 4 rupees/kilo while in another in the same region it was more than 10 rupees/kilo for the same type and quality of fish on the same day. Fish were rotting in one market even as there was a shortage in another. Typically the difference in prices across markets was around 8 rupees/kilo. After cell service came to the region, the difference dropped to less than 2 rupees/kilo.

Advances in technology are a mixed blessing, as they transform the world in which we live.[4] They have led to the most rapid decrease in extreme poverty that the world has ever seen. But they have also displaced labor and increased the premium to higher education, thus increasing inequality. Technological changes are also allowing people to form and maintain relationships at greater distances and to be increasingly selective in whom they choose to be friends with. Understanding how trends in technology affect our networks requires understanding how and why those networks form.

## Network Formation

Examining how networks form is what first got me hooked on studying networks. Why do we end up in different positions in networks? Do we form the "right" networks—ones that are best for our community or society?[5]

We have already discussed some important aspects of network formation, from the fact that centrality begets centrality, to the many forces that push us to form networks with strong homophily. But there is still another aspect of how networks form that is fundamental to understanding the consequences of technological forces on network formation.

Externalities have played a central role in all of our discussions thus far: from vaccinations to financial contagions to social learning. They make networks both interesting and important to understand. So it should come as no surprise that externalities are key to understanding network formation.

Many of the externalities that play a role in network formation are positive ones. I will start with a personal example. To help pay for college, throughout high school and college, I worked various jobs. Most were low skilled—I got them because they required almost no training and I was available. I worked as a stock boy in a local store, I worked at a warehouse unloading trucks, I worked as a night watchman at a local medical clinic, I filed papers at a local bank. The pay was generally minimum wage and I did not have any more skills when I left the jobs than when I started them. The summer before my senior year in college that changed. My undergraduate advisor knew a former colleague who was heading up a research team at the Chicago Mercantile Exchange—a futures market. My advisor's connection got me an interview and ultimately a job. Not only was the pay better, but I learned an enormous amount in a short time. That experience of seeing how markets work, grow, and change, as well as where they fall short, was important in my decision to become an economist. This was a hugely positive externality: my advisor's friendship with a former colleague was critical in my getting the job and much more.

Now, let's rewind many years. Was my advisor's decision to form that friendship with his colleague at all influenced by the chance that, years later, it would make a difference to me? Of course not.

The point is that our relationships result in information and access that can be valuable to others around us, and yet those positive benefits to others play at most a minor role in our decisions to form and maintain relationships. For example, as we saw with social learning, the more varied and wider-ranging our connections are, the faster our communities learn and the more informed they become. However, when we form friendships we don't take all of that into account. When I go to a conference or read a book, it is generally because I am interested in the topic and want to learn more about it, not necessarily because I anticipate that I might eventually pass that knowledge along to someone else. If we all were choosing our relationships in order to improve communication and knowledge in our communities we would engage with more people.

This is an example of a more general tension that exists between individual incentives to form relationships and what is best for a society, the subject of my first research on networks together with economist Asher Wolinsky. The rampant externalities in network settings make it common for people to end up forming networks that are suboptimal: people are not accounting for the full impact of their relationships and in many situations *everyone* would be made better off with a different network.[6]

This applies not only to how many relationships and interactions people have, but also to how diverse they are. People who go against tendencies of homophily can help overcome some of the barriers that we have seen homophily impose. They enrich the informational content of their community. But bridging gaps and interacting with people who are fundamentally different can be both time-consuming and difficult, resulting in a general tendency in network formation. When there are positive externalities from having more connections, people tend to underinvest in the quantity and diversity of their relationships. If we all formed more relationships, and more diverse relationships, this would enhance both the richness of information in our society and its circulation, and we would all benefit.[7]

By the same token, with negative externalities just the opposite can happen—people have more relationships than would be socially optimal. An easy example is the frequency with which people have unprotected sexual interactions, thereby transmitting STDs. Having more unprotected sex with more partners does not just endanger direct partners, but as we have seen in Chapter 3, it also increases the conductivity of the network—which has a multiplicative effect on contagion.

With these basic tendencies in mind, the impact of technology comes more clearly into focus. There are two major ways in which technology changes incentives to form relationships.[8] First, it makes it easier to form relationships and to maintain them at a distance. This is generally good news for situations with positive externalities: technology helps people form more relationships than they would otherwise. Given that people tend to underinvest in social relationships, anything that greases the wheel of interaction can be helpful.

But there is another impact of technology that comes with a downside. Advances in technology make it easier to connect with people and organizations with views similar to one's own. This can lead to greater homophily. It can be wonderful, as in the story of Dave and Liisa, but it also can lead to more fragmentation in our networks. Niche media outlets and news services that cater to specific interests and views can operate, biasing information flows. People can shift their relationships to form tighter echo chambers engaging more with like-minded others who reinforce narrow perspectives.[9]

Let's have a look at some of these forces in action. I will start with economic globalization: technology has made international trade easier, and we have a much denser global network than ever before. This has created a positive externality: the world is far more peaceful now than at any time in its history.

## Trading Networks: Make Trade Not War

*"The secret of politics? Make a good treaty with Russia."*

—OTTO VON BISMARCK

Otto von Bismarck—after wars with Denmark, Austria, and France ended in 1871—managed to unite German states around Prussia. He then started on an early attempt at globalization, with a stated goal of forming alliances to maintain peace. His triple alliance with Austria-Hungary and Italy, and nonaggression pact with Russia, led to a brief period of relative peace, at least in Europe, which unraveled and eventually erupted in the extreme violence of the First World War.

It was not until after the savagery of the Second World War that peace, by historical standards, would emerge. The average chance that two countries were at war in any given year over the last seven decades was ten times less than it was in the nineteenth and first half of the twentieth centuries! In spite of wars in Korea, Vietnam, Afghanistan, Congo, Kuwait, and Iraq, among others, the incidence of interstate war has dropped dramatically by almost any measure.

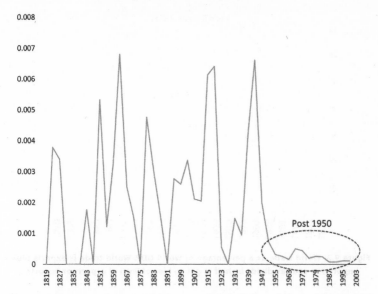

**Figure 9.1: Wars per pair of countries by year, 1820–2000, from Jackson and Nei (2015).**

The average number of wars per pair of countries per year from 1820 to 1949 was .00059, while from 1950 to 2000 it was .00006. This drop by a factor of 10 is pictured in Figure 9.1.[10]

So what accounts for the dramatic increase in peace in our lifetime? This is a question that I investigated with economist Stephen Nei, a former student of mine.

The alliances made by Bismarck and many of those up through the Second World War, especially those between potential foes, were alliances of convenience. They were not cemented by deeper economic ties. International trade was growing, but most countries still had very few trading partners on whom their economies depended.

International trade grew exponentially through the second half of the twentieth century, not just in absolute terms but actually as a fraction of the world's total production. The history of international trade is pictured in Figure 9.2.[11]

Trade grew as many countries' incomes rose, and burgeoning middle classes could afford a wider range of goods. It was greatly facilitated by falling shipping costs resulting from changes in maritime technology and the expansion of ports that could handle container shipping.

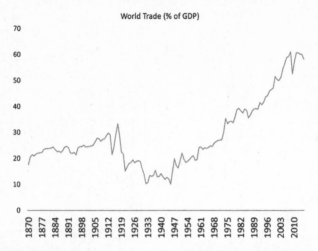

**Figure 9.2: World trade as a percentage of world GDP. World trade is exports plus imports, so you can divide by half to get an idea of level of exports or imports.**

**Table 9.1: Average number of trading partners per country. At least one country has to trade at least .5 percent of its GDP with the other for the two of them to be considered trade partners (respectively .1 percent). From Jackson and Nei (2015).**

| YEAR | 1870 | 1913 | 1950 | 1973 |
|------|------|------|------|------|
| Partner Defined ≥ .5% | 2.8 | 10.1 | 10.3 | 17.0 |
| Partner Defined ≥ 1% | 3.6 | 14.2 | 20.3 | 34.0 |

It is not just the total amount of trade that has grown, but more important countries now each have many more major trading partners, and their arrangements are more likely to last a decade. Table 9.1 shows how many trading partners countries have on average and how that has increased over time, from Bismarck's era through the early 1970s. The number has grown even more rapidly since the 1970s.

The key is that trade networks and alliance networks have moved largely in tandem.

If we look from the end of the Napoleonic Wars (starting with 1816) through 1950, each country averaged 2.5 alliances at a time. Moreover, those alliances were ephemeral. Any given alliance at any moment had just over a two-thirds chance of existing five years later. Most were marriages of convenience, not anchored by trade, and were easily disrupted when times got tough.[12] After 1951, the number of alliances per country grew by a factor of more than 4 to 10.5. Even more important, these alliances were quite steady: any given alliance now had a 95 percent chance of still existing five years later.[13] Figure 9.3 pictures what some of these networks looked like at key points in time.

Both trade and military alliances became much denser and more stable over time, coinciding with the dramatic decrease in wars. Moreover, most wars that have occurred in the past three decades have involved low-trade countries on at least one if not both sides of the conflict. Major trade partners simply do not appear on opposite sides of a war.

Countries with an extra ten trading partners are half as likely to be at war at any given moment; and pairs of countries that had a one standard deviation higher level of trade with each other have a 17 times lower chance of being at war with each other than the average pair.[14]

We cannot be sure that the striking correlations between trade and peace are causal. We don't have controlled experiments with parallel worlds, one with trade and another without, to see what would happen. There are also many other things that have changed over time: the international political landscape, the number of democracies, and the presence of nuclear weapons. All of these changes are likely contributors to our more peaceful world. But as I found together with Stephen Nei,[15] the timing of conflicts and other details suggests

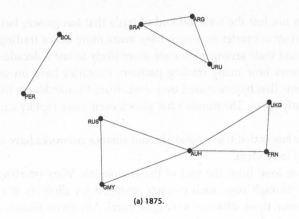

**(a) 1875.**

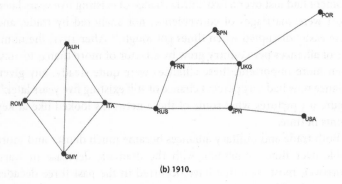

**(b) 1910.**

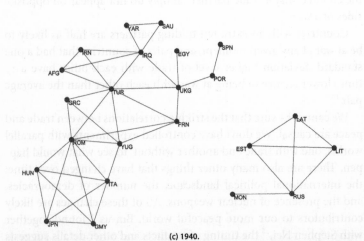

**(c) 1940.**

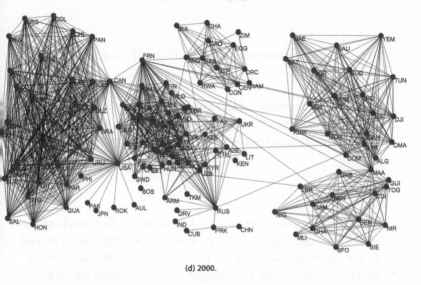

(d) 2000.

**Figure 9.3: The evolution of military alliances: We see the dramatic increase in the density of the recent network compared to historical ones.**

that none of these other factors can really explain all of what we have seen, and it seems that denser trade networks are playing a major role in generating peace.

Given the importance of world peace, even if we cannot be certain of causation, it is worth knowing the correlations and using our common sense. Two countries that have close economic ties have much less incentive to fight each other, and more incentive to protect each other. The European Union has led to the most peaceful time in Europe's history by a huge margin.

If one wants a recipe for lowering the incidence of wars in Africa and the Middle East, the message is clear: grow the economies and the regional trade networks, and especially promote trade between potential adversaries. It is not an easy task, but it is the obvious one. Signing repeated treaties, even with powerful third parties but without serious trade, has been tenuous at best.

As we have seen dangerous cycles of isolationist tendencies over the centuries, this lesson about trade networks and peace might be one of the more important ones in these pages.

## Wired: Increasing Homophily and Polarization

*"Globalization means we have to re-examine some of our ideas, and look at ideas from other countries, from other cultures, and open ourselves to them. And that's not comfortable for the average person."*

—HERBIE HANCOCK

There are signs that the increased polarization that much of the world is seeing (e.g., Figure 7.4) is not coincidental, but is related to the common forces that are increasing immobility and inequality in many places and also changing communication patterns. For example, one way to see how the Internet is changing politics is to trace when the Internet became available in different places and how the politics in those places were affected.

Anja Prummer did just this and examined how the availability of the Internet in regions within the U.S. affected how polarized their politics are. Using the congressional voting patterns of a district to estimate its polarization, Prummer estimates that when a locale has gone from having no Internet connection to complete Internet availability there is a 22 percent increase in the polarization of the district's politics (all else being equal).[16]

This is not evidence that having the Internet is what increased polarization. For instance, it could be that Internet availability was spurred by changes in the economic conditions of the area, which also affects polarization.

However, other studies have taken advantage of randomization in the way that the Internet became available in various places (for instance, due to changes in old rules governing right-of-way that affected where optic cable could be laid, or due to delays in regulatory changes that affected when the Internet made it to different regions). Studies based on these sorts of plausibly random factors affecting Internet rollout find significant evidence that the relationship is causal. For instance, Yphtach Lelkes, Gaurav Sood, and Shanto Iyengar (2015) find that increasing Internet access increases the following of partisan news and leads to more polarized politics.[17] Samuele Poy and Simone Schüller (2016) find that increased Internet availability significantly increased voter turnout and changed relative party shares, in a similar study of Internet deployment in Trento, Italy.

Polarization of beliefs is complex, and it would be wrong to blame only Internet access. Another technological force may also have played an important role. Cable television led to a proliferation of news sources some decades before the Internet came into being, and around the world people have been subject to a tug-of-war from differing ideologies for some decades.[18] The advent of cable enabled more targeted and ideologically based news outlets to reach people more cheaply at greater distances, and to reach populations that could sustain them.[19] This interacts with homophily—if a community of people all consume news from the same sources, and then talk to each other—the combination of limited inputs and echoes leads to stronger consensus within communities, and more divergent views across communities. Again, the homophily that so strongly defines our networks is a vital part of such polarization.

As we saw in Chapter 7, there is a difference between how quickly things like memes, ideas, and rumors can spread, and yet how slowly a person's beliefs about complex and subjective topics change. This suggests that the sorts of changes that come with increased connectivity and homophily can energize people with already strong views rather than dramatically change people's views. People gain confidence when others share similar views, and also tend to overestimate how many other people share their view.[20] The increased homophily can lead groups to deepen their resolve.

As we have seen, the more general consequences of increased homophily are wide and deep, including immobility and inequality.

## Disruptions

Beyond increased connectivity and homophily, changes in networks can also be disruptive, as norms and cultures are ripped apart.

The last hundred years have seen the most rapid transition from rural to urban population in human history. Two centuries ago, only 3 percent of the world's population was urban. By a century ago, it had risen to 15 percent, and by 1950 it made it to 30 percent. It hit 50 percent in 2008, and now sits at more than 54 percent, and rising. On its current trend, it will be above two thirds by mid-century.

Although that trend has slowed in parts of Europe and North America, partly because rural areas have little population left, it is accelerating in much of Asia and Africa.

It is not hard to figure out why this is happening: urban areas are where the jobs are. As agriculture shrinks as a contributor to the world economy, the advantages to living in proximity with many others when producing goods and services lead populations to agglomerate.

The relationship between how productive a country is and how urban it is is remarkable. Figure 9.4 shows the effect for a selection of countries so that their names are readable. The graph for all countries in the world is similar, just more crowded. The correlation between the productivity of a country and the urbanization rate is 72 percent over the data of all 179 countries in the data set.[21]

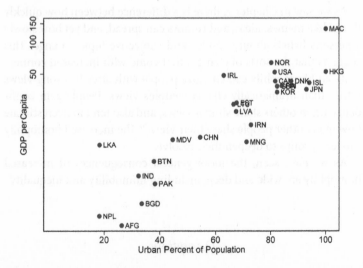

Figure 9.4: The relation between urbanization and productivity: Selected countries from Asia, Europe, and North America. Great Britain and France lie on top of each other and so are difficult to discern. The GDP per capita is in thousands of dollars and the scale is logarithmic. Data from the World Development Index of the World Bank for 2014.

Of course, this does not mean that simply building cities will make a country richer. New and rapidly growing cities often come with slums and poverty.[22] But this correlation does tell us that the inevitable shifts in economies around the world are likely to be accompanied by increased urbanization.[23]

Urban and rural populations have different network structures. In many rural areas around the world, people are often connected in tight communities, and rely on themselves, family, and friends for many if not most of their needs. With urbanization and increased market access, many needs are met through transactions rather than favors and cooperative production. This runs the spectrum from people growing their own food, to borrowing money.

A few decades ago, more than three quarters of all Chinese lived in the countryside. Now more than half live in urban areas. As the economy has shifted, so has the population. In addition to those living in cities, there are many who live in the countryside and work in the cities.[24] Such dramatic changes have torn many families apart.

For the hundreds of millions who follow Confucius, filial piety is one of the most important virtues. *The Twenty-four Filial Exemplars* is a classic text of folk stories by Guo Jujing from the Yuan dynasty that dramatize how children should act toward their elders—from fighting a tiger to save a father, to feeling a mother's pain in one's heart, to giving up a job to search for a lost mother. Here is one of the twenty-four:

> W'ang Xi'ang of the Jin dynasty lost his mother early. His stepmother, named Zhu, was unloving toward him and unceasingly spoke ill of him before his father. Because of this he lost the love of his father also. His stepmother often liked to eat fresh fish, but one winter the cold froze [the river] to ice. Xi'ang loosened his clothes and lay on the ice to [melt it so he could] procure them. Suddenly the ice opened of itself and a pair of carp leapt out. He took them and returned to serve them to his stepmother.[25]

With a growing aging population, many of whom have no pensions, support from children has become more important than ever. As the disruption in families has continued to rise, the state has even intervened. Various laws for respect of the elderly have been on the books for years, but they have recently been renewed and are now being enforced. For instance, the updated "Law of the People's Republic of China on the Protection of the Rights and Interests of the Elderly"[26] includes: "Article XIII: Dependents shall fulfill the economic support for the elderly, the obligations of life care and spiritual comfort, care for the elderly's special needs."

In the wake of the law, there have been lawsuits. For instance, a few months after the new version of the law, a seventy-seven-year-old woman sued her daughter for neglecting her duties after the two had a fight. The court ruled in the mother's favor, deciding not only that the daughter and her husband had to help support the mother financially, but also that the daughter had to visit her mother at least once every two months. Those might not be the most fun visits, but the message was clear.

We have seen similar effects in the Indian villages we studied.[27] Out of the seventy-five villages for which we mapped social networks, between 2007 and 2010, microfinance was introduced to forty-three but not the other thirty-two. The villages that got microfinance saw roughly an extra 15 percent drop in the density of their networks compared to the villages that did not get the microfinance. People started borrowing from the bank and less from their friends. This also had widespread spillovers. The advice network was similarly diminished. The spillover also affected people who did not participate in microfinance within the villages in which they could have gotten loans. Even connections between two people who decided not to participate in microfinance tended to disappear. So not only did the market exposure replace some relationships that we would expect it to, but it also led to an overall decrease in social activity within the villages.

These seismic shifts in networks, beyond the incremental ones, can be very disruptive when examining societies that have long relied on social and family ties for much of their support and culture. This is not to say that progress should stop. But it does say that large changes in where and/or how people live can upend networks, which can lead to difficulties for those who still rely on informal relationships for much of their support.

## The More Things Change

Technology will continue to advance and reshape our networks. Humans have been rewired many times: by the printing press, letter writing, trains, the telegraph, overseas travel, the telephone, the Internet, and the advent of social media. Perhaps it is our arrogance that leads us to assume that the current changes in our lives are truly revolutionary and unique.

Nonetheless, the changes are real—as we saw with networks of trade and military alliances. Still, humans are predictable. Human networks are easily recognizable and have exhibited regularities for some time. In our social transformations we see "more" of many things: denser networks, more homophily, more polarization, faster movement of information and contagion. And, as we have learned, even small changes in connectedness can have big consequences. Even changes in the extent, rather than the shape, of our networks can have profound implications—for contagion, immobility, and polarization.

There is much to be optimistic about. Economic productivity is at a historic high, large parts of the world are emerging from extreme poverty, and wars are down. People are living longer and enjoying themselves more. But there are challenges ahead. Divisions in our networks of interaction and communication remain, and some are even growing despite our advances in the ability to connect at a distance. The resulting frustration with increased immobility and inequality, when combined with growing polarization, can be volatile.

We need to better understand the benefits and risks from ever-increasing financial entanglements and address those risks rather than ignoring them. We need to recognize that our network is highly connected when it comes to the transmission of diseases, including some that we have yet to encounter and that may spread more rapidly and widely than in our worst nightmares. We need to recognize the numerous externalities in our lives, as well as how networks shape our social norms and behaviors, including corruption and crime.

We need to combat the damaging side-effects of homophily, as well as improve the incentives to collect and spread accurate and deep information, while learning to better filter the noise. Understanding the human network can ensure that our increased connectivity will improve our collective intelligence and productivity, instead of dividing society even more.

# ACKNOWLEDGMENTS

As we have seen in these pages, people are embedded in many different networks that combine to shape our views and actions. This is never truer than when writing a book. Stress also reveals much about your true friendships, and writing a book shines a bright light on relationships.

The seed for this book came from my wife's line-by-line proofreading of a text I wrote called *Social and Economic Networks*. Sara loved the descriptions of the concepts and ideas that came at the beginning of each chapter, before the technical presentation. Compiling those descriptions would have made for a short and dry book, but it made it clear that the many wonders of how networks influence human behavior could be made accessible to a wide readership, and there was at least one person interested in learning more.

A special thanks to Asher Wolinsky, for a lunch conversation in 1992 that sparked my interest in networks, and for his collaboration that launched me on the journey that eventually led to this book. I have a long list of coauthors on social interactions whose deep influences on my views are explicit or implicit in the pages here. In chronological order: Asher Wolinsky, Alison Watts, Bhaskar Dutta, Ehud Kalai, Anna Bogomolnaia, Anne van den Nouweland, Toni Calvó-Armengol, Roland Fryer, Francis Bloch, Gary Charness, Alan Kirman, Jernej Čopič, Brian Rogers, Massimo Morelli, Dunia López Pintado, Leeat Yariv, Andrea Galeotti, Sanjeev Goyal, Fernando Vega-Redondo, Ben Golub, Sergio Currarini, Paolo Pin, Daron Acemoglu, Tomas Rodriguez Barraquer, Xu Tan, Abhijit Banerjee, Arun Chandrasekhar, Esther Duflo, Yiqing Xing, Yves Zenou, Matt Elliott, Stephen Nei, Matt Leduc, Ramesh Johari, Sylvia Morelli, Desmond Ong, Rucha Makati, Jamil Zaki, Pietro Tebaldi, Mohammad Akbarpour, Evan Storms, Nathan Canen, Francesco Trebbi, Zafer Kanik, and Sharon Shiao.

I am grateful to have had much of my research, which provided the background for this book, supported by a number of organizations: Northwestern University, Caltech, Stanford University, the National Science Foundation, the Guggenheim Foundation, the Center for Advanced Studies in the Behavioral Sciences, the Army Research Office, the Canadian Institute for Advanced Research, and the Santa Fe Institute. In addition, I wish to give a shout-out to Wikipedia—it gives one faith in our human network and its ability to harvest our collective knowledge. It is unparalleled in its centrality, quickly connecting one to otherwise hard-to-find sources on almost any subject.

I have drawn on a lot of social capital in the writing of this book. My family—my wife, Sara; my daughters, Lisa and Emilie; my parents, Sally and Hal; and my brother and sister, Mark and Kim—commented on many drafts, kept my morale high, and put up with outlandish hours. My academic "parents," Hugo Sonnenschein, Salvador Barberà, and Darrell Duffie, have shaped me over the years, and their decades of friendship and mentoring are greatly appreciated. Salvador also provided very helpful comments on a draft, as did several of my academic "children," Yiqing Xing, Eduardo Laguna Műggenburg, Isa Chaves, Sharon Shiao, and Evan Storms. My literary family—Tim Sullivan, Max Brockman, Erroll McDonald, and Nicholas Thomson—guided me and kept me moving through the long process of writing such a book. Erroll McDonald is a remarkable editor and his surgical cuts and suggestions are deeply appreciated.

# NOTES

## 1. INTRODUCTION: NETWORKS AND HUMAN BEHAVIOR

1. See Brummitt, Barnett, and D'Souza (2015) for more on the role of social media and the connections across the countries in the Arab Spring.

2. This quote appears widely, and is usually attributed to Confucius. However there is no evidence of it appearing in his writings. Yiqing Xing pointed me to the closest variant of the quote in Chinese, which actually comes from the *New Books of Tang* AD 1040, which Yiqing translated as "Though peace reigns over the land, the stupid people create trouble for themselves." In any case, the misattributed quote, from wherever it originates, captures how simple things can easily become complex—a theme here.

3. The math shorthand for this is $2^{435}$, which is approximately the same as $10^{131}$.

4. Estimates of atoms in the known universe (within roughly 90 billion light-years of us) are on the order of $2^{275}$.

5. This network is discussed in more detail in Chapter 5. The data are from the Add Health Study (the National Longitudinal Study of Adolescent to Adult Health, a program project directed by Kathleen Mullan Harris and designed by J. Richard Udry, Peter S. Bearman, and Kathleen Mullan Harris at the University of North Carolina at Chapel Hill, and funded by grant P01-HD31921 from the Eunice Kennedy Shriver National Institute of Child Health and Human Development, with cooperative funding from twenty-three other federal agencies and foundations. Special acknowledgment is due Ronald R. Rindfuss and Barbara Entwisle for assistance in the original design. Information on how to obtain the Add Health data files is available on the Add Health Web site (http://www.cpc.unc.edu/addhealth). No direct support was received from grant P01-HD31921 for this analysis.

6. Given the vast terrain that networks cover, there are many sources for background on networks from a variety of perspectives. Textbooks include Wasserman and Faust (1994, sociology, statistics); Watts (1999, complex systems, sociology); Diestel (2000, graph theory); Bollabás (2001, random graph theory); Goyal (2007, economics); Vega Redondo (2007, complex systems, economics); Jackson (2008a, eco-

nomics, sociology, graph theory, complex systems); Easley and Klein-berg (2010, economics, computer science, complex systems); Barabási (2016, physics, complex systems); Borgatti, Everett, Johnson (2016, sociology, data collection, analysis). There are also several excellent books that are meant for broad audiences that discuss the importance of networks in our lives, such as Barabási (2003); Watts (2004); Chris-takis and Fowler (2009); and Ferguson (2018). Here, I synthesize and distill some concepts from this vast terrain, and also cover many sub-jects and applications that have been studied but do not appear in any of these books, together with facts about some important social and economic trends. In doing so, there are also new ideas developed and emphasized. I draw heavily from my own research, which is inevitable in writing such a book, but the spotlight is on our collective knowl-edge of networks.

7. A common term among researchers describing networks of interac-tion among humans is "social networks." That term also has a wide-spread lay interpretation that has come to mean Web-based platforms and a variety of associated social media. However, as humans are a naturally interactive and codependent species, social networks are prehistoric and embody much more than modern platforms for inter-action. There is really no distinction that I will draw between human networks and social networks.

## 2. POWER AND INFLUENCE: CENTRAL POSITIONS IN NETWORKS

1. Adrija Roychowdhury, "How Mahatma Gandhi Drew Inspiration from the American Independence Struggle," *The Indian Express,* July 4, 2016.

2. Nike alone paid him $480 million during that time.

3. Here I abstract from personal characteristics and other things that might affect why a person can reach so many others, and the extent to which someone can mobilize others. There are many features that make someone an important opinion leader beyond their network position—a subject that has been studied extensively since the semi-nal study of Katz and Lazarsfeld (1955). See Rogers (1995); Valente and Pumpuang (2007); Valente (2012) for detailed references; and Gladwell (2000) for a broad discussion.

4. Although it is not highlighted in this chapter or the others that follow, it should be clear that the type of connections and network we con-sider will change depending on the context. For instance, a politician's ability to enact a piece of legislation will depend on connections to other legislators, while that same politician's ability to mobilize vot-ers also depends on connections to media, party, and staff (and their

connections). Since two people who are connected or linked to each other in a network are not always "friends" in the true sense of the word, they are often called "neighbors" in network parlance, but we will slip back and forth in our parlance. For more on the differences and interplay among various networks, see Ferguson (2018).

5.   Feld (1991).
6.   Coleman (1961).
7.   These are just two components (connected pieces) of the network. The larger network not pictured here also exhibits the same phenomenon. One hundred forty-six girls have friends (defined mutually). Of those, 80 have fewer friends than their friends on average, while 25 have the same number as their friends, and 41 have more friends than their friends.
8.   To see similar examples illustrating biased estimation of opinions, see Lerman, Yan, and Wu (2015). One can also find examples in popular blogs (e.g., see Kevin Schaul's *Washington Post* blog from October 9, 2015, "A Quick Puzzle to Tell Whether You Know What People Are Thinking"); and experiments by Kearns, Judd, Tan, and Wortman (2009).
9.   Perkins, Meilman, Leichliter, Cashin, and Presley (1999).
10.  See Table 2 in Perkins, Haines, and Rice (2005).
11.  Valente, Unger, and Johnson (2005).
12.  Tucker et al. (2013). See also Eom and Jo (2014) for a look at some other cases where popularity correlates with some characteristic or behavior.
13.  See Jackson (2016) for a much more extensive discussion of these two effects.
14.  With careful observation one can estimate the size of the feedback effect: how different an individual's behavior is due to the fact that it takes place in a social context. For a variety of socially influenced activities these effects are more than 2 (e.g., see Glaeser, Sacerdote, and Scheinkman [2003]). In Chapter 8 we will come back to discuss these effects in a range of behaviors from criminal behavior and evading taxes to staying in school and seeking employment.
15.  Hodas, Kooti, and Lerman (2013).
16.  Frederick (2012).
17.  There are variations of this quote attributed to different people. For instance, a close variant "If you torture the data long enough, it will confess to anything" is often attributed to Darrell Huff and his 1954 book *How to Lie with Statistics*. It does not appear in that text, although he may have said it, and certainly his book makes clear why it is true and why this quote is powerful. The important point from our perspective is how multifaceted most things are, and how different they can look from alternative angles.

18. If one does not know this reference, it relates to a series of skits on the comedy show *Saturday Night Live*, in which archetypal Chicago sports fans debate the relative godliness of various Chicago sports heroes, who, of course, could defeat any non-Chicago sports (or other) persona.

19. For an interesting insight into the value of such comparisons have a look at Matisse's *Blue Nude (Memory of Biskra)*, which pushed Picasso and influenced his *Les Demoiselles d'Avignon*. Sebastian Smee's *The Art of the Rivalry* provides an illuminating account of their rivalry and the interaction between their art.

20. There is a slight difference, in that now we may double count things—as one of Ella's seven friends is Nanci, and so we count her as her own second-degree friend. The double counting actually makes the math a bit easier, as we just need to track the connections at each step and don't have to remember which ones we have already visited. The impact of double counting is discussed in Banerjee, Chandrasekhar, Duflo, and Jackson (2013, 2015).

21. If you are wondering about the scale of the numbers in the figure, the squared centrality numbers sum to one, so the vector of centrality numbers is normalized in a standard mathematical sense (according to the $L2$ or Euclidean norm).

22. Their algorithm also included some occasional random hops to a new node to restart the whole process to make sure it did not get stuck in some local neighborhood of pages that all just point to each other.

23. Search engines, including Google's, have evolved rapidly as both computing power and our experiences with the Web have advanced. Such engines now incorporate much richer information about both the user and various Web pages, and more tailored network information about how people tend to navigate the Web and what they might really be searching for. Moreover, the Web is a moving target these days, with constantly evolving content. Nonetheless, conceptually the idea of PageRank was an important breakthrough, and one that nicely embodies network information.

24. Tracking connections beyond the first degree is important in many settings beyond search engines and spreading information. Variations on such iterative calculations for centrality, and using eigenvectors, were floating around in the social networks literature for several decades before Google came into existence, with pioneering work in the 1950s by Leo Katz, and later work formalizing such methods by Phil Bonacich in the 1970s. Variations on eigenvector centrality have been used in finding "key players" in networks of illegal activities, as crime has a social component: people learn from each other about illegal opportunities and encourage each other to engage in crime, and the most central people in such networks exert the most influence

on others' participation (e.g., see Lindquist and Zenou [2014]). Such centrality measures have also been used in studies of communication among investors to predict which investors earn the highest returns in a stock market (see Ozsoylev, Walden, Yavuz, and Bildik [2014]).

25. See Karlan and Valdivia (2011), for instance, for more on how impactful simple training can be in such settings.

26. I should be careful not to paint too rosy a picture, as with any large-scale innovation that works well, one can also find horror stories. There are reports from around the world of people who borrowed too much and were financially and personally ruined, similar to people getting into excessive credit card debt and declaring personal bankruptcy. The high repayment rates in microfinance have also led many companies to enter the business, some with aggressive tactics and less ethical intentions—again, not unlike the credit card and small loan businesses in the developed world. There has also been debate about the extent to which microfinance has led to improvements in the lives of the borrowers. Sizable increases in productivity may take years to emerge and are hard to disentangle from other trends and programs, and so despite the enormous spread of microfinance throughout the world, there is mixed evidence regarding whether there are significant increases in the wealth or incomes of those gaining access to microfinance (e.g., see Banerjee and Duflo [2014]; Crépon, Devoto, Duflo, and Parienté [2015]). Nonetheless, access to credit can still help people to smooth their consumption and expenditures, which can be extremely valuable to the very poor.

27. See Schaner (2015), for example, for some discussion about how the differences in how money comes into a household can affect how it is spent.

28. Self-help groups are informal groups within villages that often hold a joint bank account and/or engage in making regular payments into the group and then getting various forms of rotating payments or loans out.

29. In contrast, in large cities such as New York, London, Sydney, or Beijing, people may interact with some people in person, others via social media, and others via phone, and their acquaintances may be placed all over the city and the larger world. Getting a fairly complete picture of such a large and diverse network is almost impossible, while in the small villages capturing the networks was manageable. We measured networks via a series of questions about the different ways in which households tend to interact with each other: borrowing and lending money, providing advice, borrowing and lending kerosene (for cooking and heating), providing help in an emergency, and so forth. The villages averaged just over two hundred households each, and each household interacted with about 15 others on average when account-

ing for all of these forms of interaction, with significant variation across households (with nontrivial numbers having a degree of less than 10 as well as above 20).

30.  In fact, a straight plot of eventual microfinance participation against the degree centrality of the initial seeds shows a slightly negative (but insignificant) slope. After controlling for a number of characteristics of the villages, one still finds no significant relationship. Details on the study can be found in the Table S3 of the supplement to Banerjee, Chandrasekhar, Duflo, and Jackson (2013).

31.  Popularity of these seeds might have been much more important if this were a new product that people were uncertain about and wanted to see what others were doing before deciding themselves. For such situations, see studies by Cai, de Janvry, and Sadoulet (2015); and Kim et al. (2015).

32.  There are many other things that might influence participation, including a friend's decision to participate. In our statistical analysis of the data we carefully control for these factors (see Banerjee, Chandrasekhar, Duflo, and Jackson [2013] for details and techniques).

33.  See the discussion in Banerjee, Chandrasekhar, Duflo, and Jackson (2015) for details on the calculations, some of which I am skipping here.

34.  By allowing for the additional flexibility of estimating the rate of communication and the number of iterations, this kind of measure should be expected to outperform the other measures. But it turns out to do better in these villages even when one fixes the rate of communication and number of iterations before analyzing the diffusion. In order not to give diffusion centrality an advantage over the other measures, we fixed the frequency of interactions between households based on some basic network characteristics—setting them just at the threshold at which information has a chance to reach everyone in the network, and the number of iterations based on the amount of time that the people were exposed to microfinance during our study. Even doing that, diffusion centrality outperforms the other measures (see column [10] of Table S3 in the supplementary materials of Banerjee, Chandrasekhar, Duflo, and Jackson [2013]). The additional variation comes from comparing the R-squares at the bottom of that table— a measure of what fraction of the variation of microfinance participation across villages is explained by the various centrality measures (together with some controls). In fact, the marginal gain in R-square that diffusion centrality adds compared to what eigenvector centrality adds is more than a factor of three (e.g., Panel C, column (2)'s R-sq. minus column (3)'s is .173, versus column (4)'s minus column (3)'s is .055, where column (3)'s provides the fit with controls and degree centrality, and degree centrality accounts for almost nothing).

35. The term "political lieutenant" is borrowed from Dale Kent (1978), whose careful data collection in the 1970s underlies the networks analyzed here. Her data were later extended and further analyzed in terms of network patterns by John Padgett and Christopher Ansell (1993). The data in the figures here are from a study by Ronald Breiger and Pip Pattison (1986), who made use of Padgett's data. The data include a different subset of families from Padgett and Ansell (1993). I have updated the data accordingly. This includes marriage and business connections between the Albizzi and Peruzzi, and an additional family, the Guasconi, on the side of the opposition to the Medici.

36. This is a point made by Padgett and Ansell (1993): "Medician political control was produced by network disjunctures within the elite, which the Medici alone spanned." For more on this perspective, see Burt (1992, 2000, 2005).

37. The Pucci were strong supporters of the Medici, particularly in terms of helping Cosimo and his family while he was exiled. So although the data do not include a formal business dealing or marriage between the families, there was a strong tie between them and it continued. Also, I include the Salviati and the Barbadori on the side of the Medici, given their eventual positions in the conflict. Dropping them, and including the tie between the Pucci and the Medici, would make the Medici figure even more starlike.

38. The contrast between the Medici bloc and the other oligarchs is starkly illustrated in a final showdown in late September of 1434, just before Cosimo de' Medici returned to Florence. Rinaldo degli Albizzi attempted to organize a takeover of the government, which was transitioning to one firmly leaning to the Medici. Rinaldo called on various other families, including the Strozzi, to send armed men; but the response to this call to arms flailed as the families debated among themselves. In contrast, Cosimo, from his central position—even in exile—easily rallied an armed force to protect the new Signoria.

39. It was introduced by Jac Anthonisse (1971) and by Linton Freeman (1977). For an illuminating discussion, see Ferguson's (2018) analysis of the Nixon and Ford networks.

40. Barabási and Albert (1999). See also Price (1976); Krapivsky, Redner, and Leyvraz (2000); Mitzenmacher (2004); Jackson and Rogers (2007a); Clauset, Shalizi, and Newman (2009). This was something that they had observed when looking at how many links there were to different Web pages on the Notre Dame domain of the Internet. Albert, Jeong, and Barabási (1999).

41. For instance, see Akcigit, Caicedo, Miguelez, Stantcheva, and Sterzi (2016) for evidence from patent data that co-patenting with more connected individuals (measured by how many people they have co-

patented with in the past) leads to more successful patents than co-patenting with less connected individuals.

42. Jackson and Rogers (2007a). Another possibility is that people copy others' links or mix random and preferential attachment (Kleinberg, Kumar, Raghavan, Rajagopalan, and Tomkins (1999); Kumar, Ragha-van, Rajagopalan, Sivakumar, Tomkins, and Upfal (2000); Pennock, Flake, Lawrence, Glover, and Giles (2002); Vázquez, (2003), which also leads to a similar distribution, with some differences in other network features.

43. For instance, see Fafchamps, van der Leij, and Goyal (2010), Chaney (2014), Jackson and Rogers (2007a).

44. Jackson and Rogers (2007a).

45. There are many more measures of centrality than those mentioned here. Some are conceptually similar, but involve slightly different cal-culations, while others involve other concepts. The mathematically inclined and endeavoring reader can find more background and ref-erences in Borgatti (2005); Jackson (2008a, 2017); Bloch, Jackson, and Tebaldi (2016); Jackson (2017).

46. This also relates to a measure called "closeness centrality," which keeps track of how close an individual is to others.

### 3. DIFFUSION AND CONTAGION

1. A study by Katharine Dean et al. (2010) suggests that the medieval spread of the Black Plague may have been primarily due to fleas and lice that live mostly on humans, and not so dependent upon rats and other animals. The hygiene of the day meant that such fleas and lice were abundant and could easily make their way from one host to another. The rarer modern cases of the plague are more dependent upon flea-bearing animals or close human-to-human contact, given that fewer people now live with lice and fleas regularly upon them so that it is now harder for, say, a flea to make its way directly from one human to another.

2. See Marvel, Martin, Doering, Lusseau, and Newman (2013).

3. This figure is based on data from a study by Peter Bearman, James Moody, and Katherine Stovel (2004), involving the Add Health data set (the National Longitudinal Adolescent Health data set, as refer-enced in Chapter 1). The figure differs slightly from their figure 2.

4. More precisely, a component is a part of a network in which every node can reach every other node via a path in the network, and is maximal in the sense that every link that involves any node in the component (and hence any neighbor of a node in the component) is included in the component.

5. Networks tend to have at most one giant component. Having two of them requires having many people in each of the two components. However, for those two components to be separate, it has to be that nobody in either component has any connections to the other component, which becomes very unlikely as the numbers of people in each of the two components increase. It only takes one connection across the two components to combine them into one.

6. I am glossing over an issue of timing here. Some of the relationships ended before others began, and so there are certain restrictions on which directions a disease might move in this network. That can slow contagion down, but would not necessarily eliminate it from still infecting large numbers of those in the giant component. See Johansen (2004); Wu et al. (2010); Barabási (2011); Pfitzner, Scholtes, Tessone, Garras, and Schweitzer (2013); and Akbarpour and Jackson (2018) for discussion and details.

7. The estimates of its prevalence vary widely depending on the sample of the population and the techniques used to measure and define infection, and are complicated by the fact that many people have no idea that they are infected. Estimates of the fraction of sexually active people who have been infected at one point in their lives are well above 50 percent (see Revzina and DiClemente [2005] for a meta-study).

8. For background, see Stanley (1971).

9. The Big Fifty rifle earned notoriety in a battle in 1874 between roughly thirty buffalo hunters and several hundred Comanche, Cheyenne, and Kiowa warriors at a trading post known as Adobe Walls in the Texas Panhandle. On the third day, one of the hunters, Billy Dixon, in what he acknowledged as a lucky shot, killed an Indian chief at an estimated 1,538 yards, helping to convince the Indians to end the battle.

10. Althaus (2014).

11. These sorts of high-level calculations are abstracting from a great amount of heterogeneity in populations—it may be that reproduction numbers within schools are much higher than in the general population. These sorts of broad estimates are made from observing the number of cases over time in large populations, and much more detailed information can be used in designing policies for stemming contagions. However, for our discussion, these high-level numbers provide the essential insights.

12. The reproduction number of a disease can be greater than one in part of a population or in some locations and not in others, and this can still lead it to reach large portions of the population and cross boundaries. A detailed analysis of this phenomenon appears in Jackson and López Pintado (2013).

13.   Sidgwick found externalities inescapable when trying to devise measures of a society's well-being. Earlier philosophers, such as John Stuart Mill and Jeremy Bentham, grappled with externalities as they developed ways of measuring a society's well-being, but failed to articulate them as clearly. Economists such as Adam Smith (1776) and Alfred Marshall (1890) mentioned the issue in their writings on the efficiency of markets, but largely sidestepped externalities, even though Marshall was familiar with Sidgwick and his writings. Marshall seems to have been distrustful of a government's ability to do much of anything, and so it is perhaps not surprising that he avoided the topic of externalities, since overcoming externalities often involves regulations, taxes, or other government interventions. Arthur Cecil Pigou is the name that many economists associate with externalities, as he mentioned them directly in a 1920 essay. Interestingly, this may be in part responding to a critique by Allyn Young, who in 1913 pointed out that such effects were glossed over in earlier work by Pigou (*Wealth and Welfare*, by A. C. Pigou, M.A., London: Macmillan, 1912) and suggested that it deserves a fuller treatment (see Young's p. 676—and thanks to Ken Arrow for pointing me to Young's paper). Yet it would not actually be until the 1960s, with a pair of papers by Ronald Coase and James Buchanan and Craig Stubblebine, that externalities would really be completely laid out in their modern form (Coase [1960]; Buchanan and Stubblebine [1962]). One can also find intermediate works that wrestle with these concepts such as that of Frank Knight (1924) and Tibor Scitovsky (1954). An excellent discussion is given by Kenneth Arrow (1969).

14.   This definition of externalities is a broadly encompassing and modern version. It does not require that the consequences of one's behaviors on others be intended, and it applies to all sorts of behaviors from a single person smoking to a tire factory polluting. It includes both positive externalities, such as someone coming up with the idea for public-key encryption for Web security, and negative externalities, such as someone cheating in a sports competition. Often externalities are incidental and not the reason for the original behavior: as in the case of a person smoking. But there are cases in which externalities are intended, as for instance when someone writes software and makes it freely available. This makes the definition a bit slippery, since one person punching another is not really what we mean to capture with the concept of externality, but it is admitted under the definition here. I will stick with the definition that may be over-inclusive for the sake of simplicity and to cover the wide variety of ways that externalities appear in networks.

15.   There are many forms of externalities in network settings. They should not be confused with the special class that are termed "network exter-

nalities." Those refer to situations in which a person's consumption value of, say, a new technology depends on how many others also use the same technology. Network externalities certainly matter in networks, but there are many other forms of externality that are of interest to us.

16. Pew Research Center Internet Project Survey, August 7–September 16, 2013, http://www.pewresearch.org/fact-tank/2014/02/03/what-people -like-dislike-about-facebook/.

17. Ugander, Karrer, Backstrom, and Marlow (2011).

18. Backstrom, Boldi, Rosa, Ugander, and Vigna (2012).

19. It was first studied mathematically by Ray Solomonoff and Anatol Rapoport (1951), and then studied more extensively in the late 1950s and early 1960s by the mathematicians Paul Erdős and Alfréd Rényi (1959; 1960), when they built the foundations of random graph theory. It is less well known that it was independently studied by Edgar Gilbert (1959).

20. More detailed numbers appear in a follow-up study by Travers and Milgram (1969).

21. See Dodds, Muhamad, and Watts (2003).

22. The answer to this question is important to cover here, but given its extensive treatment elsewhere I will stick to the basic insights. See Watts (1999) for more discussion and illustrations, and Jackson (2008) for mathematical detail.

23. More generally, beyond Facebook, estimates of the number of people known by a given individual (where known means having some contact in the last two years and reciprocally being able to contact each other) vary but are in the range of upper hundreds to thousands depending on the technique and populations used for estimation. See McCarty, Killworth, Bernard, and Johnsen (2001) and McCormick, Salganik, and Zheng (2010).

24. This is roughly right when accounting for the friendship paradox, which leads to more friends, and the rate at which Diana's friends might be friends with each other, which lowers the number of new people reached. For those of you interested in the mathematical details, there are two problems with this quick calculation. This first is that we overestimated the rate at which the neighborhoods expand, since not every friend at every step is "new": some have already been reached. For instance, some of a user's friends' friends are that user's friends. For instance, if Diana is friends with Emilie and Lisa, then when going out on paths of length 2, if they are friends with each other, then we have already counted them among the 200 friends, and so they should not be counted at the next step. Thus, the 200 new friends at each step for each person generally involves some double counting. However, even if we do a conservative approximation and

cut the number of new friends reached at each step after the first in half, to 100, we would end up with $200 \times 100 \times 100 \times 100 = 200$ million after four steps. (A better approximation has the number of new friends be higher at early steps and lower at later steps. But it generally has mostly new friends until the very last step, since most of the population is reached at the last step.) By the fifth step we would have reached 720 million users, and this tells us almost exactly why Ugander Karrer, Backstrom, and Marlow (2011) found an average distance between any two active users of 4.7. The other problem with our calculation is to assume that every friend brings in the same number of additional friends on the next step. There is substantial heterogeneity in the population and one friend might reach 500 new friends and another almost none. However, with this large a network and average degree, this variation essentially washes out. This is a fact that traces back to the work of Erdős and Rényi mentioned above (1959; 1960) for networks in which links are formed uniformly at random, and has now been established in much richer random network models (e.g., see Jackson [2008b]). Variations of laws of large numbers apply and working with approximations that ignore individual to individual variations are very accurate in many settings, even in networks as rich and global as Facebook.

25. "A Few Things You (Probably) Don't Know About Thanksgiving," Becky Little, *National Geographic*, November 21, 2016.

26. See Marvel, Martin, Doering, Lusseau, and Newman (2013) for more comparison.

27. As Cesaretti Lobo, Bettencourt, Ortman, Smith (2016) point out, medieval cities involved social and spatial structures that share many features in common with modern cities. The medieval population, however, was much more rural than a modern population, and traveled much less. For more discussion see Ferguson (2018).

28. See Valentine (2006).

29. See Altizer et al. (2006) for more detailed discussion.

30. For instance, see Jared Diamond's (1997) illuminating description.

31. See Shulman, Shulman, and Sims (2009).

32. See Schmitt and Nordyke (2001).

33. For instance, see Randy Shilts's book *And the Band Played On* (1987).

34. Lau et al. (2017).

35. Godfrey, Moore, Nelson, and Bull (2010).

36. Image from Barbulat/Shutterstock.com (Vectorstock), under an expanded license.

37. There are many other things that might also correlate with degree here, for instance, how large a territory a tuatara has, which might also relate to the animal's chance of having ticks. One cannot rule out all such alternative explanations without a controlled experiment, but

the fact that the mite infections do not exhibit the correlation with degree but are correlated with territory and body size is reassuring.

38. See Godfrey (2013).

39. Christakis and Fowler (2010).

40. Having high degree does not always mean bad news in terms of infection. Japanese macaques (monkeys) with more social contact have been found to have fewer lice than other macaques who had less social contact. This is a seasonal effect, as found in Duboscq, Romano, Sueur, and MacIntosh (2016). A key form of contact between macaques is to groom each other, with one removing lice eggs from the other—a form of true friendship. Thus, contact not only leads to the spread of lice, but also to their removal. Higher degree translates into more grooming by others and ultimately to having fewer lice.

41. Bajardi et al. (2011).

42. Cowling et al. (2010).

43. See Donald G. McNeil Jr., "In Reaction to Zika Outbreak, Echoes of Polio Global Health," *New York Times*, August 29, 2016.

44. E.g., see Ferguson et al. (2006).

45. The Super Bowl is usually the most viewed event in the U.S. each year: more than one third of the U.S. population watched its television broadcast in 2016. Although it has some international viewership, it is small in comparison to FIFA's World Cup Final or the opening of the Olympics (Beijing's opening holds the record), which can be orders of magnitude higher in viewers. Thus, it is really a U.S.-based game.

## 4. TOO CONNECTED TO FAIL: FINANCIAL NETWORKS

1. More than thirty countries are major producers of coffee (producing more than thirty million pounds per year). Global chains such as Starbucks and Costa Coffee ride out a political crisis or weather disaster that causes temporary shortages in one region by sourcing from somewhere else. World commodity prices can still be volatile (and coffee prices are certainly no exception), but a company trying to consistently deliver coffee to consumers is better off if it can buy from many countries than if it is locked into just one region and exposed to the idiosyncratic production gyrations in that one region.

2. This is just part of Sheila Ramos's story, as reported by Paul Kiel. The fuller story is fascinatingly told by Kiel in "The Great American Foreclosure Story: The Struggle for Justice and a Place to Call Home," ProPublica, 2012.

3. Fannie Mae is the common nickname for FNMA, which is the Federal National Mortgage Association; and Freddie Mac stands for FHLMC—which is the Federal Home Loan Mortgage Corporation.

Fannie Mae and Freddie Mac collected loans from all over the U.S. and packaged them into bunches called mortgage-backed securities and more generally collateralized-debt obligations. A third enterprise that was a major player in the mortgage market, Ginnie Mae (for GNMA—the Government National Mortgage Association), was government-run and specialized in loans such as those to veterans or low-income housing and other government programs. It is also large and had its difficulties, but is less central to our story.

4. Credit default swaps are the primary form of insurance that various financial institutions were buying to make sure they would be paid if a collateralized-debt obligation (e.g., a mortgage-backed security) failed to make its payments. For instance, if someone was buying a mortgage-backed security from Lehman Brothers, they might also go to AIG to buy some insurance against the once seemingly absurd event that Lehman Brothers would fail to pay on its obligation.

5. For a detailed discussion of this point, see Elliott, Golub, and Jackson (2014).

6. Beneficial trading can involve short-term borrowing and lending with other banks as their deposits fluctuate, as well as exchanging currencies as a bank's exposure to specific currencies fluctuates, and even much longer-term co-investing as banks may each have regional expertise that biases their portfolios of loans and other investments to be local. You might ask why one cares whether a bank's portfolio is unbalanced. Someone who buys shares of different banks could do the diversifying by holding a portfolio of banks even if the banks themselves are not well-diversified in their investments. Indeed, a lack of diversification by the banks themselves would not be a problem without the frictions and costs of potential bankruptcies. However, bankruptcy costs are real losses to the economy: valuable resources remain idle or underutilized while the bankruptcy is sorted out, and enormous amounts of time and legal costs are burned in the process. Those costs are what make us want to avoid any firm becoming insolvent, and also what makes it especially costly to have bankruptcies become contagious.

7. This does not mean that (c) is a "safe" situation: here each bank still only has between four and six connections, and so there can still be nontrivial exposure and a large-scale default by several banks at the same time could take others with it.

8. Guilt by association can also accompany this phenomenon. When Lehman Brothers hit bankruptcy, companies that looked like Lehman, not just ones that did business with them, were also suddenly cast into uncertainty. In fact, banks all over the world curtailed the short-term (overnight) lending among themselves, which they normally use to meet capital reserve requirements (amounts of cash that they need

to have on hand by regulation) and to do other short-term smooth-
ing. Nobody was sure which banks were safe and which were not. As
overnight trading evaporated, those markets had to be propped up by
countries' central banks.

9. For instance, see Bloom (2009).

10. For more discussion, see, for instance, Shiller (2015).

11. Prominent people can end up causing panics by their statements. If
someone who is observed by many others says that he or she has lost
confidence in some bank or other financial enterprise, even if one
does not believe them, one might fear that some of the many others
who observe the same prominent person will. For more on the ability
of prominent people to correlate the behavior of others, see Acemoglu
and Jackson (2014).

12. Branch (2002).

13. Of the lost 44 percent, 16 percent are legal costs and other losses associated with the bankruptcy itself, and the other 28 percent are losses
in the values of the assets (e.g., investments gone bad) that either trig-
gered the insolvency or came along with trying to unwind and sell off
the assets. At least the 16 percent is loss due to bankruptcy, and likely
some of the other 28 percent is due to the bankruptcy as well, as a
result of a firesale of assets.

14. See https://www5.fdic.gov/hsob/HSOBRpt.asp.

15. The arguments for and against the separation hinge on trade-offs
of conflicts of interest versus economies of scale and scope. Having
companies trying to make money by trading and selling securities,
and at the same time managing other people's money, for instance,
could lead them to push people toward securities that the company
has interests in, or to gain advantageous information about intended
buy and sell orders before other investors. The benefits of economies
of scale and scope are discussed below.

16. For more background and references on core-periphery financial
networks, see Wang (2017), Farboodi (2017); Hugonnier, Lester, and
Weill (2014); Fricke and Lux (2014); van der Leij, in't Veld, Daan and
Hommes (2016); Gofman (2011); and Babus and Hu (2017).

17. There are many other issues with having a few banks that are doing
most of the business that I will not discuss here, such as the substantial market power in determining prices and extracting profits that
comes with a few firms dominating an industry, as well as temptations
to take advantage of their ability to cross market-making and trading
boundaries.

18. See Uzzi (1997).

19. The arrows actually also incorporate indirect exposures, via a measure
that we outline in the paper (see Elliott, Golub, and Jackson [2014] for
details).

20. For instance, see the discussion of the Asian Crisis in Baig and Gold-fajn (1999). For more background on financial crises and contagions across countries, see Reinhart and Rogoff (2009) as well as Kindel-berger (2000); Neal and Weidenmier (2003); and Kaminsky, Reinhart, and Vegh (2003).

21. Battiston et al. (2016) note a related challenge in monitoring financial networks: the phase transitions and abrupt cascades that accompany financial markets mean that being slightly off in the measurement of someone's exposure to some investment, or in the network, can lead to a substantial underestimation of the potential cascade that can ensue.

22. See Admati and Hellwig (2013); Admati (2016) for further discussions of incentive problems in financial markets.

23. See Lucas (2013) for some background history.

24. In addition to the challenge of regulating moving targets, placing any restrictions on what any given financial company can do with its money has other problems too. By restricting investments, one can push the regulated enterprise to less-diversified portfolios. Many European banks are subject to some constraints on what they can invest in, and some ended up holding a large amount of Greek sovereign debt as that was allowed and offering relatively high interest rates. Regulations and restrictions on investments can also correlate investments, as the regulated entities are pushed to invest in the same more limited options, leading to many insolvencies at the same time if those limited options have a negative shock.

25. The real popping started in 2007 and continued through 2008. It began with mortgage, subprime lending, and construction companies becoming insolvent: American Freedom Mortgage, Mortgage Lend-ers Network USA, DR Horton, Countrywide Financial, New Century Financial, American Home Mortgage Investment Corp., Ameri-quest . . . it continued with banks and investment banks becoming insolvent: BNP Paribas, Northern Rock, IndyMac, Bear Stearns . . . and then Fannie Mae and Freddie Mac are taken over; and quickly more banks and investment banks are sold off or bankrupt: Meryl Lynch, Lehman Brothers, Washington Mutual; and in various install-ments AIG gets large amounts of government aid. In the fall of 2008, Congress eventually passed the $700 billion Troubled Asset Relief Program under which troubled bank assets are purchased at inflated values, and finally in November the U.S. Federal Reserve pledged another $800 billion to buy more assets, with more than half of that buying bad mortgages from Fannie Mae and Freddie Mac.

26. There are also forces that correlate the investments of various finan-cial institutions, and banks especially. Some are regulated to only be able to hold some types of assets, so they can be constrained in

their portfolios (see note 24). Also, they are often in competition with each other—if one's investments outperform another's then they can attract more customers and investors. This can lead them to copy each other—by holding similar portfolios banks can stabilize their customer base, as those customers are then more or less indifferent between them. There is also a herd mentality, which can lead them to move toward the same investments (more on this in Chapter 7, but also see Scharfstein and Stein [1990] and Froot, Scharfstein and Stein [1992]).

27. January 2011, official government edition.

28. Bank of America, Barclays, BNP Paribas, Citigroup, Deutsche Bank, Goldman Sachs, HSBC, Morgan Stanley, Royal Bank of Scotland, Société Générale, among others.

29. Six members of the commission were appointed by Democrats and four by Republicans and, not surprisingly, they disagreed on the role of regulations—or the lack thereof—in the causes of the crisis, along with various details concerning responsibilities for the crisis. Nonetheless, they agreed on the fact that potential contagion was central to the crisis and an important driver of the government's intervention. For instance, the dissenting report (page 432), in recounting the importance of the government takeover of Fannie Mae and Freddie Mac, states: "A bank cannot hold all of its assets in debt issued by General Electric or AT&T, but can hold it all in Fannie or Freddie debt. The same is true for many other investors in the United States and around the world—they assumed that GSE [government-sponsored enterprise: Fannie Mae and Freddie Mac] debt was perfectly safe and so they weighted it too heavily in their portfolios. Policymakers were convinced that this counterparty risk faced by many financial institutions meant that any write-down of GSE debt would trigger a chain of failures throughout the financial system. In addition, GSE debt was used as collateral in short-term lending markets, and by extension, their failure would have led to a sudden massive contraction of credit beyond what did occur."

## 5. HOMOPHILY: HOUSES DIVIDED

1. Some jatis do not fit cleanly within a caste. For instance, a clan that had historically included both artisans and soldiers straddles two varnas, but it was shoehorned into one caste in the nineteenth century.

2. India Human Development Survey (IHDS), conducted by the National Council for Applied Economic Research (NCAER) and the University of Maryland: http://ihds.info/.

3. See Maurizio Mazzocco and Shiv Saini (2012). Getting accurate estimates of household as well as village-wide incomes and consumptions can often be challenging (e.g., see Ravallion and Chaudhuri [1997]),

but even with noisy data, full risk sharing on local levels is still generally rejected (e.g., Townsend [1994]; Fafchamps and Lund [2003]; Kinnan and Townsend [2012]; Samphantharak and Townsend [2018]). Atila Ambrus, Markus Mobius, and Adam Szeidl (2014) look deeply into the specific network structures to understand the mechanics of such risk sharing, and offer intuitive explanations. The dense connections within groups, for instance, allow villagers to help each other out to override small idiosyncratic shocks. When large shocks hit, especially hitting a whole group, then they need to rely on relationships with other groups for help. There, the lack of connections across groups becomes inhibiting, since there are few relationships between the groups: not enough social capital exists across caste lines for them to help each other out in dealing with large hits to their incomes. This is not unique to Indian villages, as there is also ample evidence that the poorest in the United States have similar difficulties in dealing with shocks to their incomes (see, for example, Blundell, Pistaferri, and Preston [2008]). Xing (2016) explores why such inefficient risk sharing can be quite entrenched.

4. Hispanics and other categories form a small minority of this school, and in this case they end up more integrated. The level of integration varies with group size, and tiny groups sometimes have no choice but to integrate. See Currarini, Jackson, and Pin (2009, 2010).

5. The same patterns are seen when looking at recent Facebook friendships among Dutch teenagers. A study led by Bas Hofstra (Hofstra, Corten, van Tubergen, and Ellisond [2016]) finds stronger homophily among core groups of friends than overall, and finds stronger homophily on ethnicity than gender.

6. Fryer (2007).

7. The asymmetry in these numbers is due to the fact that the white population is much larger than the black population. The patterns also differ by gender, as black male and white female couples are much more common than the reverse.

8. See McPherson, Smith-Lovin, and Cook (2001) for overall background of the vast literature. A few more recent references can be found in Jackson, Rogers, and Zenou (2017). For recent studies on the genetic similarities of friends, see Christakis and Fowler (2014) and Domingue, Belsky, Fletcher, Conley, Boardman, Harris (2018).

9. Verbrugge (1977). She examined 240 different categories (for instance, ten-year groupings of age, various religions, professions, etc.). Then she calculated odds ratios for the closest friendship. For instance, she compared the odds that someone who is Catholic named a best friend who is Catholic to the odds that someone who is not Catholic named a Catholic as a best friend. If the ratio is 1, then there is no homophily:

people are naming Catholics as best friends in the same way regardless of whether they are Catholic or not. The odds ratio for Catholic were 6.3 in Detroit and 6.8 in Alt Neustadt—so Catholics were more than 6 times relatively more likely to name Catholics as best friends than non-Catholics were. Out of the 240 categories she examined, she found all odds ratios to be above 1, and 225 of the 240 odds ratios were statistically significant (the 15 insignificant categories were subsamples that were small), and those odds ratios ranged from 2.2 to 81. The extreme of 81 corresponded to an age category: when asked to name a best friend, someone younger than 25 years old was 81 times relatively more likely to name someone younger than 25 than was someone older than 25.

10. This is the East African portion, and the full rift also runs up to Lebanon. It also touches parts of Malawi, Burundi, Uganda, Rwanda, and Zambia.

11. Apicella et al. (2012).

12. For instance, see Chiappori, Salanié, and Weiss (2017) for increasing homophily trends in marriage by education level.

13. Pew Research, 2016.

14. This discussion is based on the analysis of Skopek, Schulz, and Blossfeld (2010).

15. See Table 4 in Skopek, Schulz, and Blossfeld (2010). Women's response rates are 33 percent higher than normal when a man is of a similar educational standing, and 36 percent lower than normal when a man has a lower educational standing. Men's response rates are 21 percent higher to women of similar educational standing and only 9 percent lower when contacted by a woman of lower educational standing.

16. See the study by Ken-Hou Lin and Jennifer Lundquist (2013).

17. This is from the PRRI American Values Survey of 2013. See Robert P. Jones, "Self-Segregation: Why It's So Hard for Whites to Understand Ferguson," *The Atlantic*, August 21, 2014; and Christopher Ingraham, "Three Quarters of Whites Don't Have Any Non-White Friends," *Washington Post*, August 25, 2014.

18. See Figure 13 and Table 7 in Bailey, Cao, Kuchler, Stroebel, and Wong (2017).

19. For instance, see Kenney (2000); Hwang and Horowitt (2012).

20. If you are intrigued by this example, you can find an excellent Web site for richer simulations developed by Vi Hart and Nicky Case: http://ncase.me/polygons/.

21. For more on tipping points see Granovetter (1978); Rogers (1995); Gladwell (2000); Jackson and Yariv (2011).

22. See Card, Mas, and Rothstein (2008).

23. See Easterly (2009).

24. There are also others that we will defer to Chapter 8.

25. Yiqing Xing, a former student of mine, and I studied whether people's ability to predict others' behaviors depends on their country of origin (Jackson and Xing [2014]). We used populations from the U.S. and India. People were randomly matched with a partner, and they had to coordinate on splitting a sum of money—each simultaneously stating whether they wanted an equal or unequal split, and if unequal who should get the greater share. They only got the money if their requests agreed, and so it was essential to be able to predict what the other person was going to ask for. Only 34 percent of the Indian population went for the equal split, while 83 percent of the U.S. population did. Most important, people were much more accurate in predicting the play of their own population and their overall earnings were higher when matched with their own population. Showing my own bias, I had expected to find little difference across countries, as an equal split seemed so focal to me. That was typical: many errors in prediction were from predicting that the other population would act like one's own. Other researchers have also found significant differences in play across culture populations. A group of economists and anthropologists, headed by Joe Henrich (Henrich et al. [2001]), tested levels of prosocial (altruistic) behavior across a variety of different societies and countries. They found that self-interested play was higher in cultures with less market access, and that access to markets actually led to more prosocial behavior. Their explanation for this is that prosociality and cooperation are needed to make markets work.

26. The fact that friends act similarly has not been lost on businesses. Kreditech is a company out of Hamburg that offers small loans in parts of Europe. They ask prospective borrowers for access to their social media accounts. If they see that an applicant's friend has defaulted on a Kreditech loan, then they are more likely to reject the applicant, while if they see that the applicant's friends are well-employed then they are more likely to offer a loan. ("Lenders Are Turning to Social Media to Assess Borrowers," *The Economist*, February 9, 2013).

27. Christofer Edling, Jens Rydgren, and Rickard Sandell (2016) examined millions of Spanish register records that tracked migrants' locations as well as internal movements before and after the bombings. They found a reluctance of native Spaniards to live in immigrant neighborhoods as well as a trend for Arabs to move closer to each other. Edling, Rydgren, and Sandell use a measure of segregation that varies from 0 to 1 depending on how segregated a group is—with 1 representing complete segregation and 0 being total integration. The segregation of the Arab population was near .7 in the year 2000, and had dropped to .58 by the time of the bombing. It quickly reversed

and by 2006 it was up to almost .63 again, and would take several years to move back to the pre-bombing trend.

28. See http://www.prisonexp.org/the-story/. See also, Haney, Banks, and Zimbardo (1973) for a detailed description of the protocols of the experiment.

29. This is also described in Haney, Banks, and Zimbardo (1973). A consultant on the experiment, Carlo Prescott, suggested in an editorial (*Stanford Daily,* April 28, 2005) that guards were encouraged to act in some of the ways that they did. Regardless, the point for our purposes is that the subjects assumed the identities in a short period of time and the division between the two groups was stark.

30. Alesina and Zhuravskaya (2011).

31. GDP is in PPP 2000 international dollars (purchasing power parity adjusted dollars). If you wish to work with the data, it is quite illuminating and you can find the file I used to create these figures on my Web site—but credit for the data should go to (and details on precise variable definitions can be found in) Alesina and Zhuravskaya (2011).

32. You may be questioning whether segregation is the real driver of the poor functioning of a country or just a by-product. It could be that poorly functioning countries somehow encourage people to segregate, or remain separated, by ethnicity. There is nothing in the figure that shows any causal relationship. We cannot do a carefully controlled experiment to test such a theory, since we cannot create some segregated countries and some nonsegregated countries, and then put them in identical situations and watch them evolve over time. Despite not being able to do an experiment, Alberto Alesina and Ekaterina Zhuravskaya (2011) provide evidence suggesting that these relationships are causal. They use a technique that is common for dealing with such data, using "instrumental variables." The ideas behind this technique, and its application here, are explained in an online supplementary appendix to this text that can be found on my Web site, under Chapter 5.

33. It also helps us navigate a network. You can find more discussion in the online supplementary appendix and also in Chapter 5 in Duncan Watts (2004).

## 6. IMMOBILITY AND INEQUALITY: NETWORK FEEDBACK AND POVERTY TRAPS

1. For those of you who do not remember Olympia Beer's publicity, or who thought Artesia was a place, artesian water refers to water that is under natural pressure, and often sitting in an aquifer that is at an angle and that provides water with little effort.

2. Claire Vaye Watkins, "The Ivy League Was Another Planet," *New York Times,* March 28, 2013.

3. See https://www.theodysseyonline.com/@alonaking.

4. There are many ways of measuring mobility—how children's education, income, wealth, social class, longevity all depend on those attributes of their parents, and one can measure father-son, mother-daughter, mother-son, etc. The enormous research on such measures finds similar patterns to what we see above—significant correlations and a lot of variation across countries; although there can be some differences in the magnitudes depending on what exactly is being measured. The term "elasticity" is what economists use to refer to things that are all done in percentage terms, in order to compare across situations. The numbers here are calculated by estimating the following sort of equation: $ln(Incomechild) = \alpha + \beta ln(Incomeparent) + error,$ and so $Incomechild = factor \times Incomeparent.$ The numbers are similar for sons and daughters. See Lee and Solon (2009).

5. This is capturing the correlation in income across generations, rather than wealth. Inheritance of wealth provides a direct channel of correlation of wealth across generations, while the correlation of income stems from many things other than wealth transmission.

6. See "The American Dream? Social Mobility and Equality in the US & EU," Dalia Research, March 1, 2017. A detailed study by Alesina, Stantcheva, and Teso (2018) finds similar overoptimism in the U.S. compared to a series of European countries. They also find that the overoptimism is highest in some of the states in which mobility is lowest. Moreover, they find that people's perceptions of mobility correlate with their political views, perceptions of government, and support for redistribution.

7. You might notice that the name is a bit off for this graph. Jay Gatsby did in fact manage to become very wealthy, albeit as a bootlegger, and hence was economically quite mobile, despite his background. It was a particular social ladder that he failed to climb. Nonetheless, the name is catchy, and does evoke the immobility in the U.S., which now goes beyond social class to include the economic.

8. For the correlation in Denmark see the supplementary material of Calvó-Armengol and Jackson (2009), while the U.S. correlation comes from Huang (2013). The attainment numbers are 43 percent for the U.S. and 39 percent for Denmark, from http://www.russell sage.org/research/chartbook/percentage-population-select-countries -bachelors-degrees-or-higher-age (accessed December 6, 2016).

9. If you want to see more detail, an online appendix on my Web site gives two other ways of thinking of Gini indices, as well as some other measures of inequality.

10. There are other ways to understand and visualize the Gini index, and the interested reader can find more detail in the online appendix. I learned the method explained here from Sam Bowles, and it is much simpler than the usual explanations, which appear in the appendix.

11. In many countries getting accurate income information for the full population is challenging. Figure 6.3 involves a series of different studies and depends on local government reporting and so should be interpreted with some caution. Moreover, Gini indices depend on whether one accounts for taxes and welfare programs. If a country has a high tax rate on high incomes and then offers large subsidies to the poor, then whether one looks at pre-tax or post-tax incomes can make a difference. In some European countries looking at post-tax rather than pre-tax incomes lowers the Gini significantly. The Gini index for the United States drops by about 10 percentage points when including taxes and transfers. Moreover, by reducing a complex phenomenon to one number, the Gini index does not give us a full picture of whether inequality is due only to the lower class being extremely poor or whether the middle class is also relatively poor. Nonetheless, these numbers do give us a rough idea of how unequally income is distributed.

12. See Lindert and Williamson (2012).

13. Haden (1995); Hayden (1997).

14. Bowles, Smith, and Borgerhoff Mulder (2010).

15. Although inequality has been increasing within many countries throughout the world, it has been decreasing between countries, as key developing countries are getting wealthier overall even as inequality grows within them. See the online appendix for more background.

16. See Goldin and Katz (2009) for an insightful overview of the evolution of education and productivity through the twentieth century.

17. Acemoglu et al. (2016) have a direct estimate of 10 percent, but then accounting for further equilibrium effects (from lost business in other parts of the economy that spill over) doubles the estimate. Similarly, Hicks and Devaraj (2015) estimate that 87 percent of the decrease in manufacturing jobs has been from changes in technology and only 13 percent from production leaving the U.S.

18. Parts of information technology are lumped into the service sector, but people have suggested expanding the number of sectors that track the production of the economy to accommodate the richer range of modern economic activities.

19. See Acemoglu and Autor (2011), page 1051, as well as Goldin and Katz (2009) for more details.

20. See Piketty (2014), Tables 7.2 and 7.3.

21. See Krugman (2014).

22. See Kaplan and Rauh (2013) for evidence that the rising wages among the top one percent are driven by technological changes that have greatly increased the productivity and scale at the top, rather than changing norms or manipulations of what executives should earn. The top one percent includes millions of people in a variety of different jobs.

23. In particular, for families who are in the bottom fifth of the income distribution, less than 30 percent of their children end up with college degrees. In contrast, families in the top fifth of the income distribution have more than 80 percent of their children end up with college degrees. Chetty and Hendren (2015).

24. American Academy of Arts and Sciences, "A Primer on the College Student Journey," 2016.

25. More than 70 percent of students at the most competitive universities come from the wealthiest quarter of the population, while only 3 percent come from the poorest quarter of the population. From a report of the Jack Kent Cooke Foundation: Giancola and Kahlenberg (2016).

26. Chetty, Friedman, Saez, Turner, and Yagan (2017).

27. This is measured ten years after graduation, for students who had federal financial aid.

28. Hart and Risley (1995).

29. Differences are not just in speaking to young children. Flavio Cunha analyzed how many hours mothers spend with their children doing a variety of things: "soothing the baby when he/she is upset; moving the baby's arms and legs around playfully; talking to the baby; playing peek-a-boo with the baby; singing songs with the baby; telling stories to the baby; reading books to the baby; and taking the baby outside to play in the yard, park, or playground." Mothers spend, typically, somewhere around four or five hours per day interacting with their young children in these ways. This varies, however, with the mothers' background. Mothers who have at least a two-year college degree spend more than two thirds of an hour a day more than do mothers who were high school dropouts with their young children. See Cunha (2016).

30. Heckman (2012). For more background and evidence on the longer-term effects of early interventions, and complementarities, see Currie (2001); Garces, Thomas, and Currie (2002); Aizer and Cunha (2012); and Felfe and Lalive (2018).

31. Boneva and Rauh (2015), Figure 4.

32. Boneva and Rauh (2015), Figure 5.

33. Hoxby and Avery (2013), Figure 2.

34. Again, see Giancola and Kahlenberg (2016).

35. Cunha, Heckman, Lochner, and Masterov (2006, page 703). See also Carneiro and Heckman (2002).

36. This is among students who are dependent upon their families, from Figure O in American Academy of Arts and Sciences, "A Primer on the College Student Journey," 2016.

37. Claire Vaye Watkins, "The Ivy League Was Another Planet."

38. See Jackson (2017) for more background on definitions of social capital.

39. We are skipping some forms of capital. For instance, "somatic capital" refers to the health and strength of one's body, which depends on calories, exercise, genetics, access to health care, and so forth. This is important in some settings, for instance, in a society of hunter-gatherers, but is less vital in our discussion. We are also skipping "cultural capital" (Bourdieu and Passeron [1970]), which becomes slippery to define, as culture is itself a rich concept with many interpretations. Our discussion will not require making cultural capital explicit, although it will be implicit.

40. Social capital has been defined in many ways, and appears sporadically over the last century. See, for instance, the many definitions in Dasgupta and Serageldin (2001) and Sobel (2002), and the discussion in Jackson (2017). Two of the first to define social capital explicitly in terms similar to the way we use it here were Pierre Bourdieu and Glenn Loury. Pierre Bourdieu defined it as (Bourdieu, 1986) "the aggregate of the actual or potential resources which are linked to possession of a durable network of more or less institutionalized relationships of mutual acquaintance or recognition." Glenn Loury states (Loury, 1977), "It may thus be useful to employ a concept of 'social capital' to represent the consequences of social position in facilitating acquisition of the standard human capital characteristics." Our definition goes beyond Loury's in allowing social capital to be used for more than acquiring human capital—for instance, calling on a friend for a loan. For details on alternative sources and measures of social capital, see Jackson (2017).

41. In terms of our discussion of centrality measures, concepts like eigenvector and diffusion centrality can capture aspects of social capital that will be missed by degree centrality.

42. 2104 American Community Survey, U.S. Census Bureau.

43. Bischoff and Reardon (2014). Let us call a family "poor" if its income is less than two thirds of the median family income, a family "'rich" if its income is more than three halves of the median family income. Define a neighborhood as poor if the majority of families in the neighborhood are poor, and call a neighborhood rich if the majority of families are rich. In 1970, 7 percent of families lived in poor neighborhoods, 8 percent in rich neighborhoods, and the remainder in mixed neighborhoods. In fact, 65 percent lived in neighborhoods in which the median income was between 80 and 125 percent of the U.S. median

income. By 2009, 15 percent of families lived in poor neighborhoods, 18 percent lived in rich neighborhoods, and only 42 percent lived in neighborhoods in which the median income was between 80 and 125 percent of the U.S. median income. For more detailed ways of measuring geographic segregation, see Echenique and Fryer Jr. (2007).

44. Earlier studies that looked at short-term outcomes were mixed in terms of finding significant impact beyond mental health. For instance, see Clampet-Lundquist and Massey (2008); Ludwig et al. (2008); de Souza Briggs, Popkin, and Goering (2010); Fryer Jr. and Katz (2013). The longer-term effects are now measurable and provide clear evidence of the impact of the program, as shown by Chetty and Hendren (2015); Chetty, Hendren, and Katz (2016b).

45. There can be some selection effects in these numbers, since families choose whether to move, and some did not move. One has to take into consideration effects such as having families who care more about their children be the most likely to make use of the vouchers and move. To avoid such issues, Raj Chetty and Nathan Hendren (Chetty and Hendren, 2015) compare siblings whose families moved (so there were two children of different ages whose family moved to see how different their outcomes are), among other things to check that the effects are causal and appropriately estimated.

46. This includes Pierre Bourdieu, James Coleman, Glenn Loury, Douglas Massey, Elinor Ostrom, Robert Putnam, and William Julius Wilson, among many others. The literature is too immense to summarize succinctly, but for a range of classic references from different fields and perspectives see Bourdieu and Passeron (1970); Loury (1977); Coleman and Hoffer (1987); Ostrom (1990); Massey and Denton (1993); Putnam (2000); Wilson (2012).

47. Britton et al. (2016).

48. Chetty, Friedman, Saez, Turner, and Yagan (2017), in looking at U.S. data, find slightly more of an effect in family background determining which higher education children attend.

49. As reported by *The New York Times,* January 27, 2013. The quote is from John Sullivan.

50. Kasinitz and Rosenberg (1996).

51. Ibid.

52. Yes, he's the same man who later served as secretary of state. He negotiated a key accord with China when tensions over U.S. relations with Taiwan were high, and was instrumental in convincing President Reagan to have extended talks with Mikhail Gorbachev, setting the stage for a thaw and ultimately an end to the Cold War. But before George Shultz was a politician, he was an economist at MIT and the University of Chicago in the 1950s through the early 1970s.

53. Myers and Shultz (1951).

54. Rees et al. (1970).

55. See Myers and Shultz (1951); Rees et al. (1970); Granovetter (1973); Montgomery (1991); Ioannides and Datcher-Loury (2004).

56. Calvó-Armengol and Jackson (2004, 2007, 2009); Jackson (2007).

57. The logic of getting more bids/opportunities leading to a better price is an old one in economics, driving the analysis of auctions, for instance. Its role in explaining wage differences by race is explored by Arrow and Borzekowski (2004) (see also Smith [2000]). It also, for instance, plays a role in why the disabled get worse prices for car repairs than nondisabled, as they seek fewer quotes (it is more time-consuming for them to access multiple shops to get quotes) and this is also anticipated by the repair shops who give them higher quotes, as analyzed by Gneezy, List, and Price (2012).

58. See Calvó-Armengol and Jackson (2004, 2007). There are some interesting subtleties in these settings. Friends help provide information about jobs, but they also compete for jobs, and so much of the technical detail that Toni Calvó-Armengol and I wrestled with was in sorting those effects out. This effect is present in the data. As Lori Beaman finds (Beaman, 2012), if refugees are relocated into a community of many others from their same country, then this increases their employment if those refugees have been there for a while, as they provide job information, while it hurts their employment if those others are arriving at the same time, as they compete for jobs. For more details on the fraction of co-ethnics in an area and employment, see Munshi (2003); Patacchini and Zenou (2012).

59. One theory (among several) on the term "doughboys" traces the term back to the Mexican-American War of 1846–1848. The infantry, in marching over dusty trails, accumulated dust and looked much like the adobe houses of northern Mexico. The alliteration of "adobe" became "doughboy." See http://www.worldwar1.com/dbc/originalb.htm.

60. Marmaros and Sacerdote (2002).

61. For anyone reading this who has had a paper rejected by a journal, take heart. Granovetter's "The Strength of Weak Ties" paper was first submitted to the *American Sociological Review* and it was rejected. Quotes from the two referees include: "it should not be published. I respectfully submit the following among an endless series of reasons that immediately came to mind." "I find that his scholarship is somewhat elementary. . . . [he] has confined himself to a few older and obvious items." See https://scatter.files.wordpress.com/2014/10/granovetter-rejection.pdf.

62. Granovetter (1973, 1995).

63. Bramoullé and Rogers (2010) also find less homophily for more popular students: if a student is named just once as a friend then there is a

75 percent chance it is by someone of the same gender, but if they are named by ten students then there is just a 51 percent chance of them being of the same gender. This would be consistent with the fewer friendships being stronger ties, on average, for instance.

64. See Gee, Jones, and Burke (2017) for evidence that any single strong tie is significantly more related to future employment than a weak tie.
65. Lalanne and Seabright (2016).
66. Lalanne and Seabright (2016), and see also Weichselbaumer and Winter-Ebmer (2005) and England (2017).
67. Beaman, Keleher, and Magruder (2016). See also Mengel (2015).
68. Fernandez, Castilla, and Moore (2000).
69. See, for instance, Fernandez, Castilla, and Moore (2000) and Dhillon, Iversion, and Torsvik (2013).
70. Pallais and Sands (2016).
71. Pallais and Sands also found that the referred workers performed even better when paired with their referrers on tasks that required some teamwork (see also Brown, Setren, and Topa, [2012]). They found no evidence that the referred employees worked any harder because they cared about the referrer's reputation. But other studies in which employers can actively penalize the referrer if the referred does not work out, as in a Bangladeshi garment factory (see Heath [forthcoming]), find that the referred does seem to react to reputational concerns. Also, Beaman and Magruder (2012) find that if pay is based on the referred's performance, then better people are referred.
72. Alfred died of an early attempt at an inoculation. See Susan Flantzer, "Smallpox Knew No Class Boundaries," The Site for Royal News and Discussion, http://www.unofficialroyalty.com/royal-illnesses-and-deaths/smallpox-knew-no-class-boundaries/.
73. The word "game" is not meant to trivialize the situation, as dropping out is not a game—it is terminology that comes from a branch of mathematics known as "game theory." Game theory is a tool that social scientists use to understand how people make decisions when those decisions are interconnected.
74. For more discussion, see Morris (2000); Jackson (2007, 2008); Jackson and Zenou (2014); Jackson and Storms (2017).
75. Calvó-Armengol and Jackson (2009).
76. For more details on the trends discussed here, see Abramitzky (2011, 2018).
77. To address the small portion of jobs that are being exported from developed countries, one could work at a policy that would be truly beneficial on a world scale: enforcing world work standards. Costs of labor around the world vary substantially, especially for unskilled labor—at times reaching ratios of twenty or more. Part of this has to do with varying standards and lack of enforcement of work safety,

hours, minimum wages, benefits, and general working conditions. Putting a priority on enforcing global minimal labor standards would help slow some of the trends, and likely in a much more productive manner than imposing trade barriers.

78. David Autor has an excellent TED talk on the general subject of changing workforces. "Will Automation Take Away All Our Jobs?," Ideas.Ted.com, March 29, 2017.

79. Peter S. Goodman, "Free Cash in Finland, Must Be Jobless," *New York Times*, December 17, 2016; and "Not Finnished: The Lapsing of Finland's Universal Basic Income Trial," *The Economist*, April 26, 2018.

80. The costs of such a policy of a base income are not unthinkable. For the cost of the wars in Iraq and Afghanistan (conservatively estimated at about $2.4 trillion), one could give each person in the lowest 20 percent of the income distribution in the U.S. $5,000 per year for the next eight years. For a family of four, that would be $20,000 in cash, tax free each year. If one accounts for the savings of other programs that this would offset, the net cost could be much lower. It is radical, but not infeasible.

81. Thaler and Sunstein (2008).

82. See the interview with sociologist Sean Reardon in Sam Scott, "The Gravity of Inequality," *Stanford Magazine*, December 15, 2016.

83. Nguyen (2008); Jensen (2010); Attanasio and Kaufmann (2013); Kaufmann (2014).

84. See David Leonhardt, "Make Colleges Diverse," *New York Times*, December 13, 2016.

85. See Carrell and Sacerdote (2017), for instance.

86. Moreover, centers of innovation attract further business and investment, as well as highly skilled labor forces that provide the engine of further growth. We have seen that from the urban areas of the world that are attracting investment and generating new growth: Silicon Valley, Seoul, Boston, Toronto, Taipei, Hong Kong, Singapore, Bangalore, Amsterdam, Munich, Stockholm, Shanghai. . . .

## 7. THE WISDOM AND FOLLY OF THE CROWD

1. That is, thousands of trillions.

2. For a fascinating discussion of what sets humans apart and more on our collective intelligence, see Henrich (2015).

3. For example, see Susan Pater, "How Much Does Your Animal Weigh?," University of Arizona Cooperative Extension.

4. Galton (1907); Surowiecki (2005).

5. The original article reports only the median and parts of the distribution. A letter exchange with the journal a few weeks later reports the average of 1,197.

6. For more discussion on the role of diversity in how the crowd can get things right, see Surowiecki (2005); Page (2008).

7. Group versus individual performance is the bread and butter of social psychology and there are many examples in which groups outperform individuals, including situations in which they outperform their best members at a task, and it can depend on how the team is organized. For more background and references, see Davis (1992); Hinsz, Tindale, and Vollrath (1997); Hogg and Tindale (2008).

8. This does not necessarily equal the average belief, but one that depends on the bettors' tolerance for risk. See Manski (2006); Wolfers and Zitzewitz (2006); Gjerstad (2004).

9. For more background on prediction markets see Wolfers and Zitzewitz (2004); Hahn and Tetlock (2006); Arrow et al. (2008).

10. See Roth (2007) for more on repugnant markets and this example.

11. See Page (2008, 2017) for more discussion, as well as the book by Garry Kasparov and Daniel King: *Kasparov Against the World: The Story of the Greatest Online Challenge.*

12. It turns out the same model had been earlier described by psychologist John R. P. French and briefly studied by graph theorist Frank Harary (French [1956]; Harary [1959]). DeGroot rediscovered the model and began to analyze it in more detail. An independent rediscovery often suggests a good idea.

13. For details on how this is calculated see Jackson (2008).

14. For those who want a bit more mathematical background. Two conditions are needed in the DeGroot model in order to guarantee a consensus. One is that information from any one person can flow to any other via some directed path in the network. The second is a condition known as aperiodicity: that the least common divisor of all (directed) cycles in the network is one. See Golub and Jackson (2010).

15. This presumes the eigenvector centralities have been normalized to sum to one; if not, just divide by the total sum of the eigenvector centralities.

16. Such double counting also goes by the name of "correlation neglect" and has important implications in voting and other behaviors. For instance, see Demarzo, Vayanos, and Zweibel (2003); Glaeser and Sunstein (2009); Levy and Razin (2015).

17. Chandrasekhar, Larreguy, and Xandri (2015).

18. For instance, see Choi, Gale, and Kariv (2005); Mobius, Phan, and Szeidl (2015); Enke and Zimmermann (2015); Brandts, Giritligil, and Weber (2015); Battiston and Stanca (2015).

19. Bailey, Cao, Kuchler, and Stroebel (2016).

20. For instance, see Choi, Gale, and Kariv (2005).

21. To fully account for whether a network is well-balanced, we need to adjust for how much each person interacts with others and how much

they weight those other opinions when they update their own opinions. The full description is in Golub and Jackson (2010). The speed with which beliefs converge will also depend on the network structure and those weights (Golub and Jackson [2012]).

22. For more about such advantages and disadvantages to different network structures, and opinion leaders, see Katz and Lazarsfeld (1955); Jackson and Wolinsky (1996); Bala and Goyal (1998); Golub and Jackson (2010, 2012); Galeotti and Goyal (2010); Molavi, Tahbaz-Salehi, and Jadbabaie (2018).

23. Lazarsfeld, Berelson, and Gaudet (1948).

24. Katz and Lazarsfeld (1955).

25. Gladwell (2000).

26. "Thirty Years Later: 1982 Bordeaux No Single Vintage Did More to Change the World of Wine than 1982," Wine Spectator blogs, posted March 26, 2012.

27. Andrew Edgecliffe-Johnson, "Robert Parker, the American Bacchus," *Financial Times,* December 14, 2012. Together with Nicolas Carayol, I have examined the accuracy and ratings of nineteen prominent wine experts (see Carayol and Jackson, 2017)—and the accuracies and influence have some interesting nuances.

28. McCoy (2014).

29. This abstracts away from personal tastes, which can lead to more diverse views on products like wine, restaurants, and movies. Still, despite individual variation in evaluations of some products, there can still be a sort of common base assessment that can be communicated, and the large industry of rating such goods suggests that hearing other people's opinions is valuable.

30. There are many issues with reviews. For instance, many product review systems have heavily skewed distributions of ratings—with most giving excellent scores and occasional poor scores, but few scores in the middle. Somebody who loves or hates a product may be motivated to write a review, while people who feel lukewarm are less motivated to spend time entering a review. The few reviewers who spend lots of time writing many reviews might also have biases, including being paid to like products. See Duan, Gu, and Whinston (2008); Fradkin, Grewal, Holtz, and Pearson (2015); Tadelis (2016); and Nei (2017).

31. Deer later published more details about the study and problems with the ethics and data. See, for instance, "How the Vaccine Crisis Was Meant to Make Money," *British Medical Journal,* January 11, 2011.

32. The accusations of Brian Deer came after the General Medical Council had begun its investigation of the study.

33. For an overview of some the many studies that eventually emerged on the subject, see Gerber and Offit (2009).

34. World Health Organization statistics.

35. Carrillo-Santisteve and Lopalco (2012).

36. "A lie can travel halfway around the world while the truth is still putting on its shoes" is a related quote that is often attributed to Mark Twain, but does not seem to appear in his writings. The Swift quote appeared in an article in *The Examiner* in 1710. A study by Vosoughi, Roy, and Aral (2018) provides a detailed example of this.

37. December 23, 2016, Twitter.

38. October 31, 1938.

39. See the discussion in Bartholomew (2001).

40. See http://news.bbc.co.uk/2/hi/health/2284783.stm, September 27, 2002.

41. See "Extinction of Blondes Vastly Overreported," *Washington Post*, October 2, 2002.

42. "Evaluating Information: The Cornerstone of Civic Online Reasoning," Executive Summary, Stanford History Education Group. They evaluated over 7,800 students from across twelve states.

43. Malcom Moore, "World War 2 Anniversary: The Scoop," *The Telegraph*, August 30, 2009; "Obituary: Clare Hollingworth," BBC, January 10, 2017; Margalit Fox, "Clare Hollingworth, Reporter Who Broke News of World War II, Dies at 105," *New York Times*, January 10, 2017.

44. Beyond timing issues, the production of news is undergoing a fundamental shift in how revenues are generated. The Internet poses a number of challenges for news services. Revenue from classified advertising has plummeted as has circulation of print versions of news. Revenue from online advertising, and online subscriptions, has not replaced disappearing streams of traditional revenues. For backgrounds, see for instance, Waldman (2011); Hamilton (2016); the Reuters Institute Digital News Report 2017 (http://po.st/lfJFXh); Pew Research Center "Newspaper industry estimated advertising and circulation revenue" (June 1, 2017, http://www.journalism.org/chart/newspaper-industry -estimated-advertising-and-circulation-revenue/).

45. Waldman and the Working Group on Information Needs of Communities (2011).

46. FBI press release: "Former Upper Big Branch Mine Superintendent Sentenced to Prison in Connection with a Federal Investigation at Upper Big Branch," from the U.S. Attorney's Office, Southern District of West Virginia, January 17, 2013; U.S. Department of Labor, Mine Safety and Health Administration, Coal Mine Safety and Health report of the Fatal Underground Mine Explosion April 5, 2010, Upper Big Branch Mine-South, Performance Coal Company Montcoal, Raleigh County, West Virginia, ID No. 46-08436.

47. Even after the investigation, two employees were sentenced to less than five years of prison each, and the former CEO of the company

was convicted of only one misdemeanor charge of willfully conspiring to violate health and safety standards, and sentenced to one year in prison.

48. American Society for News Editors Newsroom Employment Census projections, 1978–2014, "State of the News Media 2016," as reported by the Pew Research Center.

49. Waldman and the Working Group on Information Needs of Communities (2011), page 10.

50. The study is Kaplan and Hale (2010), and you can find a breakdown of the content of a half hour of local news in the Appendix (Section 9.4).

51. Hamilton (2016).

52. Teemu Henriksson, "Full Highlights of World Press Trends 2016 Survey," World Association of Newspapers and News Publishers.

53. Waldman and the Working Group on Information Needs of Communities (2011), page 12.

54. Gentzkow, Shapiro, and Taddy (2016).

55. See Figure 4 in ibid.

56. For other evidence, see Prummer (2016).

57. The code was developed by Renzo Lucioni, and modified by Peter Aldhous, with some adjustments for my analysis.

58. The choice of a threshold for what fraction of bills two senators had to vote the same on in order to be linked, and the year examined matters (e.g., see the Blog by Randy Olson of December 21, 2013, at http://www.randalolson.com/blog/). A threshold of one half is one where senators who are linked then agree more than they disagree. I used data as recent as possible at the time of the analysis for this chapter (2015), and back to the comparable date of 1990 from the partisanship data discussed earlier. You can find the code and data and details behind the construction of the figures on my Web site if you wish to explore the relationships in more detail.

59. In 2015 it pulls Marco Rubio and Ted Cruz, both Tea Party members, to the top, and Lindsey Graham, who battled the Tea Party, to the bottom.

60. The data that I analyze are different, but one can find the analysis that inspired this analysis on Lucioni's Web site: https://renzo.lucioni.xyz/senate-voting-relationships/. See also Moody and Mucha (2013) for other measures and similar trends.

61. See Jackson (2008b) and Section VII in Golub and Jackson (2012).

62. This presumes that the network is not completely fragmented, and that no part of the network is substantially denser than some other part. For instance, if everyone in marketing is in contact with dozens of other employees, while people in production interact only with one or two others, then this could fail, but simply because different parts

of the network have different diffusion properties, not because of the homophily.

63. This contrast is something we discuss in much more technical detail in Golub and Jackson (2012).

64. See Lena H. Sun, "Anti-Vaccine Activists Spark a State's Worst Measles Outbreak in Decades," *Washington Post,* May 5, 2017.

65. Barnes et al. (2016).

66. Barnes et al. (2016).

## 8. THE INFLUENCE OF OUR FRIENDS AND OUR LOCAL NETWORK STRUCTURES

1. See the book *The Ants,* Hölldobler and Wilson (1990).

2. Prabhakar, Dektar, and Gordon (2012).

3. See Bjorn Carey, "Stanford Researchers Discover the 'Anternet,' " *Stanford Report,* August 24, 2012.

4. For illuminating and detailed discussions of culture and evolution, see Boyd and Richerson (1988); Richerson and Boyd (2008); Tomasello (2009).

5. Coffman, Featherstone, and Kessler (2016).

6. Manski (1993); Aral, Muchnik, and Sundararajan (2009).

7. Gilchrist and Sands, 2016. If anything, this underestimates the social effect, since some people who might have wanted to see the movie during the first weekend in New York did not, so there is some pent-up demand in New York. See also Moretti (2011).

8. Lerner and Malmendier (2013).

9. Kloumann, Adamic, Kleinberg, and Wu (2015).

10. Duflo and Saez (2003).

11. Vosoughi, Roy, and Aral (2018).

12. It is difficult to test directly for such informational effects just using stock data, even though there is suggestive evidence as discussed by Bikhchandani and Sharma (2000) and Hirshleifer and Hong Teoh (2003). However, when you put people in a lab setting, so you control the information and can see what people buy and sell, then there is clear herding behavior, as seen in Anderson and Holt (1997) and Hung and Plott (2001).

13. This is another quote of dubious origin. It is not clear that Confucius ever said it, although it is widely attributed to him. It is more likely a morphing of a quote from *Analects,* a text that was likely written centuries later by followers, and has a fascinating history of its own.

14. The logic of herding comes from a pair of studies by Abhijit Banerjee (1992) and Sushil Bikhchandani, David Hirshleifer, and Ivo Welch (1992).

15.  Foster and Rosenzweig (1995). See also Conley and Udry (2010).

16.  Leduc, Jackson, and Johari (2016).

17.  Smith and Sorensen (2000); Acemoglu, Dahleh, Lobel, and Ozdaglar (2011).

18.  Elaine McArdle, "How to Eat at Sushi Dai, Tokyo: Tips and Guide to Getting a Seat at Sushi Dai, Tsukiji Fish Market," June 30, 2016, https://www.thewholeworldisaplayground.com/how-to-eat-seat -sushi-dai-tokyo-tips-guide/.

19.  It seems that fish suppliers take advantage of our inability to judge. In a study of sushi restaurants in Los Angeles (a town with a good repu- tation for sushi), a team of researchers (Willette et al. [2017]) gathered 364 fish samples from twenty-six restaurants, in a variety of neigh- borhoods, price ranges, and ratings levels, over a four-year period. Using analysis of the DNA of the fish samples, they checked whether the fish that they were served matched what it was sold as. Every single restaurant sold them at least one piece of sushi that was not as advertised. Overall, 47 percent of the sushi was mislabeled! Some species were almost always mislabeled—usually ones for which there are look-alike substitutes. For instance, 89 percent of red snapper and 93 percent of what was sold as yellowtail were something else, while only 8 percent of the mackerel was mislabeled. One cannot just blame the restaurants, as they are part of a chain of suppliers, but it is clear that somewhere along the line someone is taking repeated advantage of our inability to discern types of fish.

20.  See Young (1996) for more discussion of conventions. The driving example is an interesting one, since some countries have dramati- cally changed sides. The divisions that remain are often separated by bodies of water or major mountain ranges, with little traffic between countries.

21.  Glaeser, Sacerdote, and Scheinkman (2003); Martinelli, Parker, Pérez- Gea, and Rodrigo (2015); Su and Wu (2016).

22.  Fisman and Miguel (2007).

23.  Kuwait, with corruption of -1.1, is dropped from the figure because it is so far out of the picture (250 parking tickets per diplomat) that it squeezes the rest of the picture. It is a true outlier, not only in the huge number of tickets it had, but also in having a lower than average corruption level. (The corruption number for Kuwait in the Fisman and Miguel data seems to be out of line with many other indices that have its corruption as being close to the levels of India, Brazil, Bul- garia, and other countries with significant corruption.) I still include Kuwait in all of the numbers quoted, and, even with Kuwait included, there is a significant correlation between a country's corruption and how many tickets its diplomats amass. For instance, the comparison

of 23 and 12 includes Kuwait among noncorrupt. Dropping Kuwait from that calculation would lead to an even starker contrast, as then the average tickets among noncorrupt countries drops to 8!

24. This is also related to the concept of "transitivity": if A knows B and B knows C, does A know C?

25. Chandrasekhar and Jackson (2016).

26. See Watts (1999) for some examples, and you can find a range of clustering coefficients in data sets described in Jackson and Rogers (2007a).

27. Chandrasekhar and Jackson (2016).

28. Jackson and Rogers (2007a).

29. There are also cases in which diffusion of a new product can happen without such reinforcement. For example, we found no social reinforcement effects in the diffusion of microfinance in our Indian village study (Banerjee, Chandrasekhar, Duflo, and Jackson, 2013). There it seems that microfinance was something that villagers were already somewhat familiar with and able to decide about on their own, but that information about the availability of microfinance was something that they needed to know about. Simple awareness acts more like a disease in its spread and so does not depend on clustering or other local reinforcement to spread.

30. Goel et al. (2016).

31. See Beaman, BenYishay, Magruder, and Mobarak (2015). Damon Centola (2011) tested how diffusion worked in a setting in which people might care to have multiple friends act in order to act themselves, among an online community of more than fifteen hundred people from a health interest site. He connected some of the participants to each other in a highly clustered network and the rest of the participants were connected to each other in a network with low clustering. Each of the participants thus got some new "health buddies" and they could learn things about what their buddies were doing. He then seeded one of the people with a chance to join a health forum. Each time someone joined, his or her health buddies would get a message telling them that the buddy had joined. So if multiple buddies of someone joined the forum then they would get multiple messages. Damon found rates of diffusion of the forum that were 42 percent higher in clustered networks than in the random networks. This seemed to be a case in which multiple stimuli were important in getting people to act, and even though the clustered network had greater average distances between people, the diffusion spread faster and to more people than in the less clustered network.

32. The concept of "Nash equilibrium" is what was referred to when I mentioned multiple equilibria in settings with coordination issues. John Nash proved that such equilibria exist in a wide class of settings. It has become a primary tool in modern economics.

33. This definition of support and the game theory behind it are developed in Jackson, Rodriguez Barraquer, and Tan (2012). You might be wondering how having friends in common relates to clustering. Both involve triangles in networks and so there is a close relationship. If lots of relationships are supported, then there will be lots of triangles and so the network will tend to be highly clustered. But support and clustering are distinct conceptually. For instance, in the Indian villages, more than 90 percent of relationships are supported; however, the clustering tends to be only between 20 and 30 percent. When we think about a supported relationship, it is a relationship—a link in the network—that is what is supported by the friend in common. Clustering asks what fraction of a person's friends are friends with each other. Support and clustering also can play quite distinct roles in enforcing behavior. Support plays a role via a witness, guarantor, or enforcer, as the third party to a relationship, while clustering ensures that people can coordinate their behaviors in influencing someone else. See Coleman (1988) for more on this latter point.

34. There are times at which nuclear families require more calories than they produce. For instance, see Hill and Hurtado (2017).

35. See Kaplan, Hill, Lancaster, and Hurtado (2000) and Hill (2002), for example.

36. For an overarching perspective on human cooperation see Seabright (2010).

37. See Jackson, Rodriguez Barraquer, and Tan (2012); Feinberg, Willer, and Schultz (2014); and Ali and Miller (2016). Coleman (1988) examines a different aspect of such communication as he discusses how clustering can help two friends of a given person coordinate in sanctioning a given person, which is different from what is discussed here. See Jackson, Rodriguez Barraquer, and Tan (2012) for more discussion of the distinction.

38. Jackson, Rodriguez Barraquer, and Tan (2012).

39. The precise percentages quoted here come from the second wave of our data (that we collected in Banerjee, Chandrasekhar, Duflo, and Jackson [2016]). The rates of support were similar whether we looked at the borrowing of money, kerosene, or rice, or the offering of help or advice. Interestingly, support was significantly lower when we looked at more purely social interactions, such as having someone come to visit, which have no associated incentive issues. Also, the 93 percent support figure may be an underestimate, as when collecting network information by surveys one can miss some links—especially as not all people participate in the surveys.

40. Banerjee and Duflo (2012).

41. Breza and Chandrasekhar (2016).

42. Uzzi (1996), pp. 679–80.

43. See Jackson and Xing (2018) for a discussion of how such community enforcement can complement legal enforcement of transactions.
44. Fisman, Shi, Wang, and Xu (2018).
45. Fisman, Shi, Wang, and Xu (2017).
46. Shi and Rao (2010).
47. For some analysis of some of these competing forces, see, for instance, Bandiera, Barnakay, and Rasul (2009); and Beaman and Magruder (2012).
48. Backstrom and Kleinberg (2014).
49. See, for example, Easley and Kleinberg (2010); Uzzi (1997); and the classic article on embeddedness by Granovetter (1973).
50. For this calculation they used a recursive definition, and examined relationships that had been active for at most twelve months.

## 9. GLOBALIZATION: OUR CHANGING NETWORKS

1. Caitlin Giddings, "This Couple Found Love Across the Globe Through Strava: The Incredible Story of How a Competitive Cycling App United Two Riders from Across the World," *Bicycling Magazine*, January 25, 2016.
2. Alex Shashkevich, "Oldest Adults May Have Much to Gain from Social Technology, According to Stanford Research," *Stanford News*, November 28, 2016.
3. See Jensen (2007). The standard deviation of prices across markets over the mean price was in the 60 to 70 percent range before phones, and less than 15 percent afterward.
4. See Friedman (2016) for a wider discussion; and Hjort and Poulsen (2018) for an example of how the Internet affects productivity and employment.
5. The subject of network formation has been analyzed extensively. For our purposes here, I just focus on a couple of important insights, but there are many more in the literature. More technical and detailed discussion and numerous references can be found in Jackson (2008) and Jackson, Rogers, and Zenou (2017).
6. Jackson and Wolinsky (1996).
7. There are nuances here that I am ignoring to get across a basic insight. The way in which externalities and incentives to form relationships interact can be complex in some situations and has resulted in many studies of how networks form. You can find much more technical discussions of the literature in Jackson (2003, 2008a, 2014).
8. There is a third fairly important force that I discuss in the online appendix available on my Web site. Network formation also involves feedback. The more friends a person has, the easier it is for them to meet more people, and the more attractive they are as a friend to oth-

ers given the increased externalities that they offer. This feedback can lead to inequality in network formation: people with more friends gain even more friends.

9. As we saw in Chapter 7, for things like the formation of people's opinions, it can be the relative numbers of connections that people have with others rather than the absolute numbers that are consequential (details on this appear in Golub and Jackson [2012]). If new relationships are added that are more homophilistic, then the network becomes more homophilistic and opinions can remain or become more polarized, even as the network becomes denser.

10. The fact that there has been a significant decrease is robust to when the cutoff is made, as noted in Jackson and Nei (2015): from 1820 to 1959 it was .00056, while from 1960 to 2000 it was .00005, from 1820 to 1969 it was .00053, and from 1970 to 2000 it was .00005. In terms of wars per country (instead of per pair of countries): from 1820 to 1959 the rate was .012, while from 1960 to 2000 it was .004. Based on all "Militarized Interstate Disputes" (MID2s-MID5s) instead of just wars (MID5s—involving at least 1,000 deaths): from 1820 to 1959 there are .006 MIDs per pair of countries, while from 1960 to 2000 there were .003. Nonetheless, whether the change in conflict is statistically significant depends on what class of random processes one presumes generates the conflict. There are some processes for which it will take much more data to be sure, as pointed out by Clauset (2017).

11. Data are from 1870–1949: Klasing and Milionis (2014); 1950–1959 Penn World Trade Tables Version 8.1; 1960–2015 World Bank World Development Indicators.

12. In fact, ignoring the Second World War, the number of alliances was only 1.7 per country from 1816 to 1940.

13. Jackson and Nei (2015).

14. This is discussed in more detail in Jackson and Nei (2015). See also Li et al. (2017).

15. Jackson and Nei (2015).

16. Prummer (2016). See also Hampton and Wellman (2003); Rainie and Wellman (2014); and Sunstein (2018) for more background.

17. See also Davis and Dunaway (2016).

18. See the discussion in Gentzkow, Shapiro, and Taddy (2016); Boxell, Shapiro, and Shapiro (2017); Fiorina (2017); and Gentzkow (2017). There is still quite a bit of disagreement (polarization?) among researchers as to how to measure polarization, the extent to which it is increasing, and what the causes could be.

19. See the discussion in Gentzkow, Shapiro, and Taddy (2016).

20. This is known as the "false consensus effect" in psychology (e.g., see Ross, Greene, and House [1977]), and can be exacerbated by homoph-

ily as people's friends often influence their estimation of what others believe.

21. I thank Jasmin Droege's blog "Urbanization vs GDP per Capita Revisited," April 11, 2016, for pointing me to the data.

22. E.g., see Barnhardt, Field, and Pande (2016).

23. Ultimately, technology may also enable people to work remotely, a sort of countervailing force. But as people are increasingly dependent on others for the provision of the goods and services that they consume, there is likely to still be some agglomeration, whether it be suburban, urban, or sprawling metropolitan areas.

24. Some work and live in cities without proper permission. The "hukou" system of registering families to certain locations in China has kept the surge toward cities from becoming overwhelming, making it a challenge for some to migrate to cities.

25. Translation from David K. Jordan (1973).

26. Draft date, summer 2012.

27. Banerjee, Chandrasekhar, Duflo, and Jackson (2018).

# BIBLIOGRAPHY

Abramitzky, Ran (2011). "Lessons from the Kibbutz on the Equality-Incentives Trade-off." *The Journal of Economic Perspectives*, Vol. 25, pp. 185–207.

—— (2018). *The Mystery of the Kibbutz: Egalitarian Principles in a Capitalist World*. Princeton, N.J.: Princeton University Press.

Abramitzky, Ran, Leah Boustan, and Katherine Eriksson (2016). "Cultural Assimilation During the Age of Mass Migration." National Bureau of Economic Research Working Paper No. 22381.

Acemoglu, Daron, and David Autor (2011). "Chapter 12: Skills, Tasks and Technologies: Implications for Employment and Earnings." In *Handbook of Labor Economics*, Vol. 4B, pp. 1043–1171.

Acemoglu, Daron, David Autor, David Dorn, Gordon H. Hanson, and Brendan Price (2016). "Import Competition and the Great US Employment Sag of the 2000s." *Journal of Labor Economics*, Vol. 34, pp. S141–S198.

Acemoglu, Daron, Munther A. Dahleh, Ilan Lobel, and Asuman Ozdaglar (2011). "Bayesian Learning in Social Networks." *The Review of Economic Studies*, Vol. 78, pp. 1201–36.

Acemoglu, Daron, and Matthew O. Jackson (2014). "History, Expectations, and Leadership in the Evolution of Social Norms." *The Review of Economic Studies*, Vol. 82, pp. 423–56.

Acemoglu, Daron, Asuman Ozdaglar, and Alireza Tahbaz-Salehi (2015). "Systemic Risk and Stability in Financial Networks." *The American Economic Review*, Vol. 105, pp. 564–608.

Acemoglu, Daron, and James A. Robinson (2012). *Why Nations Fail: The Origins of Power, Prosperity, and Poverty*. New York: Crown Business.

—— (2015). "The Rise and Decline of General Laws of Capitalism." *The Journal of Economic Perspectives*, Vol. 29, pp. 3–28.

Adamic, Lada, Thomas Lento, Eytan Adar, and Pauline Ng (2014). "The Evolution of Memes on Facebook." *Facebook Data Science*, January 8.

Admati, Anat R. (2016). "It Takes a Village to Maintain a Dangerous Financial System." Available at SSRN: http://ssrn.com/abstract=2787177.

Admati, Anat R., and Martin F. Hellwig (2013). *The Banker's New Clothes: What's Wrong with Banking and What to Do About It*. Princeton, N.J.: Princeton University Press.

Agan, Amanda Y., and Sonja B. Starr (2016). "Ban the Box, Criminal Records, and Statistical Discrimination: A Field Experiment." University of Michigan Law and Economics Research Paper No. 16-012.

Agostinelli, G., J. M. Brown, and W. R. Miller (1995). "Effects of Normative Feedback on Consumption Among Heavy Drinking College Students." *Journal of Drug Education,* Vol. 25:1, pp. 31–40.

Aizer, Anna, and Flavio Cunha (2012). "The Production of Human Capital: Endowments, Investments and Fertility." National Bureau of Economic Research Working Paper No. 18429.

Akbarpour, Mohammad, and Matthew O. Jackson (2018). "Diffusion in Networks and the Virtue of Burstiness." *Proceedings of the National Academy of Sciences,* Vol. 115 (30), pp. E6996–7004.

Akbarpour, Mohammad, Suraj Malladi, and Amin Saberi (2017). "Diffusion, Seeding and the Value of Network Information." Mimeo, Stanford University.

Akcigit, Ufuk, Santiago Caicedo, Ernest Miguelez, Stefanie Stantcheva, and Valerio Sterzi (2016). "Dancing with the Stars: Interactions and Human Capital Accumulation." Mimeo, University of Chicago.

Albert, Réka, Hawoong Jeong, and Albert-László Barabási (1999). "Internet: Diameter of the World-wide Web." *Nature,* Vol. 401, pp. 130–31.

Aldrich, Daniel P. (2012). "Social, Not Physical, Infrastructure: The Critical Role of Civil Society After the 1923 Tokyo Earthquake." *Disasters,* Vol. 36, pp. 398–419.

Alesina, Alberto, Stefanie Stantcheva, and Edoardo Teso (2018). "Intergenerational Mobility and Preferences for Redistribution." *The American Economic Review,* Vol. 108, pp. 521–54.

Alesina, Alberto, and Ekaterina Zhuravskaya (2011). "Segregation and the Quality of Government in a Cross Section of Countries." *The American Economic Review,* Vol. 101, pp. 1872–1911.

Ali, Mir M., and Debra S. Dwyer (2011). "Estimating Peer Effects in Sexual Behavior Among Adolescents." *Journal of Adolescence,* Vol. 34, pp. 183–90.

Ali, S. Nageeb, and David A. Miller. (2016). "Ostracism and Forgiveness." *The American Economic Review,* Vol. 106 (8), pp. 2329–48.

Allcott, Hunt, Dean Karlan, Marcus M. Möbius, Tanya S. Rosenblat, and Adam Szeidl (2007). "Community Size and Network Closure." *The American Economic Review,* Vol. 97, pp. 80–85.

Allport, Gordon W. (1954). *The Nature of Prejudice.* Garden City, NY: Doubleday.

Althaus, Christian L. (2014). "Estimating the Reproduction Number of Ebola Virus (EBOV) During the 2014 Outbreak in West Africa." *PLOS Currents,* Vol. 2 (1), pp. 1–9.

Altizer, Sonia, Andrew Dobson, Parviez Hosseini, Peter Hudson, Mercedes Pascual, and Pejman Rohani (2006). "Seasonality and the Dynamics of Infectious Diseases." *Ecology Letters,* Vol. 9, pp. 467–84.

Altonji, Joseph G., and Rebecca M. Blank (1999). "Race and Gender in the Labor Market." *Handbook of Labor Economics*, Vol. 3, pp. 3143–3259.

Amaral, Luis A. Nunes, Antonio Scala, Marc Barthelemy, and H. Eugene Stanley (2000). "Classes of Small-World Networks." *Proceedings of the National Academy of Sciences*, Vol. 97, pp. 11149–52.

Ambrus, Attila, Markus Mobius, and Adam Szeidl (2014). "Consumption Risk-Sharing in Social Networks." *The American Economic Review*, Vol. 104, pp. 149–82.

Anderson, Katharine A. (2017). "Skill Networks and Measures of Complex Human Capital." *Proceedings of the National Academy of Sciences*, 114 (48), pp. 12720–24.

Anderson, Lisa R., and Charles A. Holt (1997). "Information Cascades in the Laboratory." *The American Economic Review*, Vol. 87, pp. 847–62.

Angrist, Joshua D., and Kevin Lang (2004). "Does School Integration Generate Peer Effects? Evidence from Boston's Metco Program." *The American Economic Review*, Vol. 94, pp. 1613–34.

Anthonisse, Jac M. (1971). "The Rush in a Directed Graph." Stichting Mathematisch Centrum, Mathematische Besliskunde, BN 9/71.

Apicella, Coren L., Frank W. Marlowe, James H. Fowler, and Nicholas A. Christakis (2012). "Social Networks and Cooperation in Hunter-Gatherers." *Nature*, Vol. 481:7382, pp. 497–501.

Aral, Sinan, Lev Muchnik, and Arun Sundararajan (2009). "Distinguishing Influence-Based Contagion from Homophily-Driven Diffusion in Dynamic Networks." *Proceedings of the National Academy of Sciences*, Vol. 106, pp. 21544–49.

—— (2013). "Engineering Social Contagions: Optimal Network Seeding in the Presence of Homophily." *Network Science*, Vol. 1, pp. 125–53.

Aral, Sinan, and Christos Nicolaides (2017). "Exercise Contagion in a Global Social Network." *Nature Communications*, Vol. 8, article no. 14753.

Arrow, Kenneth J. (1969). "The Organization of Economic Activity: Issues Pertinent to the Choice of Market Versus Nonmarket Allocation." U.S. Joint Economic Committee of Congress, 91st Congress, 1st Session. Washington, D.C.: U.S. Government Printing Office, Vol. 1, pp. 59–73.

—— (1998). "What Has Economics to Say About Racial Discrimination?" *The Journal of Economic Perspectives*, Vol. 12, pp. 91–100.

—— (2000). "Observations on Social Capital." In *Social Capital: A Multifaceted Perspective*, Vol. 6, pp. 3–5. Washington, D.C.: World Bank Publications.

Arrow, Kenneth J., and Ron Borzekowski (2004). "Limited Network Connections and the Distribution of Wages." FEDS Working Paper No. 2004-41.

Arrow, Kenneth J., Robert Forsythe, Michael Gorham, Robert Hahn, Robin Hanson, John O. Ledyard, Saul Levmore, Robert Litan, Paul Milgrom, Forrest D. Nelson, George R. Neumann, Marco Ottaviani, Thomas C.

Schelling, Robert J. Shiller, Vernon L. Smith, Erik Snowberg, Cass R. Sunstein, Paul C. Tetlock, Philip E. Tetlock, Hal R. Varian, Justin Wolfers, and Eric Zitzewitz (2008). "The Promise of Prediction Markets." *Science,* Vol. 320, pp. 877–78.

Atkeson, Andrew, and Patrick J. Kehoe (1993). "Industry Evolution and Transition: The Role of Information Capital." Federal Reserve Bank of Minneapolis, Research Department Staff Report No. 162, pp. 1–31.

Atkinson, Anthony B., and Salvatore Morelli (2014). "Chartbook of Economic Inequality." Society for the Study of Economic Inequality Working Paper No. 2014-324.

Attanasio, Orazio, and Katja Kaufmann (2013). "Educational Choices, Subjective Expectations, and Credit Constraints." National Bureau of Economic Research Working Paper No. 15087.

Auerbach, Alan J., and Kevin Hassett (2015). "Capital Taxation in the Twenty-First Century." *The American Economic Review,* Vol. 105, pp. 38–42.

Austen-Smith, David, and Roland G. Fryer, Jr. (2005). "An Economic Analysis of Acting White." *The Quarterly Journal of Economics,* Vol. 120, pp. 551–83.

Babus, Ana, and Tai-Wei Hu (2017). "Endogenous Intermediation in Over-the-Counter Markets." *Journal of Financial Economics,* Vol. 125, pp. 200–215.

Backstrom, Lars, Paolo Boldi, Marco Rosa, Johan Ugander, and Sebastiano Vigna (2012). "Four Degrees of Separation." In *Proceedings of the 4th Annual ACM Web Science Conference,* pp. 33–42.

Backstrom, Lars, and Jon Kleinberg (2014). "Romantic Partnerships and the Dispersion of Social Ties: A Network Analysis of Relationship Status on Facebook." In *Proceedings of the 17th ACM Conference on Computer Supported Cooperative Work & Social Computing,* pp. 831–41.

Badev, Anton (2017). "Discrete Games in Endogenous Networks: Theory and Policy." arXiv preprint arXiv: 1705.03137.

Baig, Taimur, and Ilan Goldfajn (1999). "Financial Market Contagion in the Asian Crisis." *IMF Staff Papers,* Vol. 46 (2), pp. 167–95.

Bailey, Mike, Rachel Cao, Theresa Kuchler, and Johannes Stroebel (2016). "The Economics Effects of Social Networks: Evidence from the Housing Market." Available at SSRN: https://ssrn.com/abstract=2753881.

Bailey, Mike, Rachel Cao, Theresa Kuchler, Johannes Stroebel, and Arlene Wong (2017). "Measuring Social Connectedness." National Bureau of Economic Research Working Paper No. 23608.

Bajardi, Paolo, Chiara Poletto, Jose J. Ramasco, Michele Tizzoni, Vittoria Colizza, and Alessandro Vespignani (2011). "Human Mobility Networks, Travel Restrictions, and the Global Spread of 2009 H1N1 Pandemic." *PLoS ONE,* Vol. 6:1, Article number 16591.

Bala, Venkatesh, and Sanjeev Goyal (1998). "Learning From Neighbours." *The Review of Economic Studies,* Vol. 65(3), pp. 595–621.

Bandiera, Oriana, Iwan Barankay, and Imran Rasul (2009). "Social Connections and Incentives in the Workplace: Evidence from Personnel Data." *Econometrica*, Vol. 77, pp. 1047–94.

Banerjee, Abhijit V. (1992). "A Simple Model of Herd Behavior." *The Quarterly Journal of Economics*, pp. 797–817.

Banerjee, Abhijit V., Arun G. Chandrasekhar, Esther Duflo, and Matthew O. Jackson (2013). "Diffusion of Microfinance." *Science*, Vol. 341 (6144), Article number 1236498, doi: 10.1126/science.1236498.

——— (2015). "Gossip and Identifying Central Individuals in a Social Network." Available at SSRN: http://ssrn.com/abstract=2425379.

——— (2018). "Changes in Social Network Structure in Response to Exposure to Formal Credit Markets." Mimeo, Stanford University.

Banerjee, Abhijit V., and Esther Duflo (2012). *Poor Economics: A Radical Rethinking of the Way to Fight Global Poverty*. New York: PublicAffairs.

——— (2014). "Do Firms Want to Borrow More? Testing Credit Constraints Using a Directed Lending Program." *The Review of Economic Studies*, Vol. 81, pp. 572–607.

Banfi, Edward C. (1958). *The Moral Basis of a Backward Society*. New York: Free Press.

Barabási, Albert-László (2003). *Linked: The New Science of Networks*. New York: Basic Books.

——— (2011). *Bursts: The Hidden Patterns Behind Everything We Do, from Your E-mail to Bloody Crusades*. New York: Penguin.

——— (2016). *Network Science*. Cambridge, U.K.: Cambridge University Press.

Barabási, Albert-László, and Réka Albert (1999). "Emergence of Scaling in Random Networks." *Science*, Vol. 286, p. 509.

Barbera, Salvador, and Matthew O. Jackson (2017). "A Model of Protests, Revolution, and Information." Available at SSRN: https://ssrn.com/abstract=2732864.

Barnes, Michele, Kolter Kalberg, Minling Pan, and PinSun Leung (2016). "When Is Brokerage Negatively Associated with Economic Benefits? Ethnic Diversity, Competition, and Common-Pool Resources." *Social Networks*, Vol. 45, pp. 55–65.

Barnes, Michele L., John Lynham, Kolter Kalberg, and PingSun Leung (2016). "Social Networks and Environmental Outcomes." *Proceedings of the National Academy of Sciences*, Vol. 113, pp. 6466–71.

Barnhardt, Sharon, Erica Field, and Rohini Pande (2016). "Moving to Opportunity or Isolation? Network Effects of a Randomized Housing Lottery in Urban India." *American Economic Journal: Applied Economics*, Vol. 9:1, pp. 1–32.

Bartholomew, Robert E. (2001). *Little Green Men, Meowing Nuns and Head-Hunting Panics: A Study of Mass Psychogenic Illness and Social Delusion*. Jefferson, NC: McFarland.

Bassok, Daphna, Jenna E. Finch, RaeHyuck Lee, Sean F. Reardon, and Jane Waldfogel (2016). "Socioeconomic Gaps in Early Childhood Experiences: 1998 to 2010." *AERA Open*, Vol. 2 (3), article no. 2332858416653924.

Battiston, Pietro, and Luca Stanca (2015). "Boundedly Rational Opinion Dynamics in Social Networks: Does Indegree Matter?" *Journal of Economic Behavior & Organization*, Vol. 119, pp. 400–421.

Battiston, Stefano, Guido Caldarelli, Robert M. May, Tarik Roukny, and Joseph E. Stiglitz (2016). "The Price of Complexity in Financial Networks." *Proceedings of the National Academy of Sciences*, 113 (36), pp. 10031–36.

Beaman, Lori A. (2012). "Social Networks and the Dynamics of Labour Market Outcomes: Evidence from Refugees Resettled in the U.S." *Review of Economic Studies*, Vol. 79:1, pp. 128–61.

Beaman, Lori A., Ariel BenYishay, Jeremy Magruder, and Ahmed Mushfiq Mobarak (2015). "Making Networks Work for Policy: Evidence from Agricultural Technology Adoption in Malawi." Working Paper, Northwestern University.

Beaman, Lori A., Niall Keleher, and Jeremy Magruder (2016). "Do Job Networks Disadvantage Women? Evidence from a Recruitment Experiment in Malawi." Unpublished, http://faculty.wcas.northwestern.edu/˜lab823 /BKM recruitment Oct2013. pdf.

Beaman, Lori A., and Jeremy Magruder (2012). "Who Gets the Job Referral? Evidence from a Social Networks Experiment." *The American Economic Review*, Vol. 102, pp. 3574–93.

Bearman, Peter S., James Moody, and Katherine Stovel (2004). "Chains of Affection: The Structure of Adolescent Romantic and Sexual Networks." *American Journal of Sociology*, Vol. 110:1, pp. 44–91.

Benhabib, Jess, Alberto Bisin, and Matthew O. Jackson (2011). *Handbook of Social Economics*. Amsterdam: North-Holland.

Benzell, Seth G., and Kevin Cooke (2016). "A Network of Thrones: Kinship and Conflict in Europe, 1495–1918." Manuscript, Boston University.

Berens, Guido, and Cees van Riel (2004). "Corporate Associations in the Academic Literature: Three Main Streams of Thought in the Reputation Measurement Literature." *Corporate Reputation Review*, Vol. 7, pp. 161–78.

Bertrand, Marianne, and Sendhil Mullainathan (2004). "Are Emily and Greg More Employable Than Lakisha and Jamal? A Field Experiment on Labor Market Discrimination." *The American Economic Review*, Vol. 94, pp. 991–1013.

Bikhchandani, Sushil, David Hirshleifer, and Ivo Welch (1992). "A Theory of Fads, Fashion, Custom, and Cultural Change as Informational Cascades." *Journal of Political Economy*, Vol. 100 (5), pp. 992–1026.

Bikhchandani, Sushil, and Sunil Sharma (2000). "Herd Behavior in Financial Markets." *IMF Economic Review*, Vol. 47, pp. 279–310.

Bischoff, Kendra, and Sean F. Reardon (2014). "Residential Segregation by Income, 1970–2009." In *Diversity and Disparities: America Enters a New Century.* New York: The Russell Sage Foundation.

Bisin, Alberto, and Thierry Verdier (2001). "The Economics of Cultural Transmission and the Dynamics of Preferences." *Journal of Economic Theory,* Vol. 97, pp. 298–319.

Blandy, Richard (1967). "Marshall on Human Capital: A Note." *Journal of Political Economy,* Vol. 75, pp. 874–75.

Blau, Peter M. (1987). "Contrasting Theoretical Perspectives." In *The Micro-Macro Link,* pp. 71–85. Berkeley: University of California Press.

Bloch, Francis, Garance Genicot, and Debraj Ray (2007). "Reciprocity in Groups and the Limits to Social Capital." *The American Economic Review,* Vol. 97, pp. 65–69.

Bloch, Francis, Matthew O. Jackson, and Pietro Tebaldi (2016). "Centrality Measures in Networks." Available at SSRN: http://ssrn.com/abstract=2749124.

Blondel, Vincent D., Jean-Loup Guillaume, Renaud Lambiotte, and Etienne Lefebvre (2008). "Fast Unfolding of Communities in Large Networks." *Journal of Statistical Mechanics: Theory and Experiment,* Vol. 100, article no. P10008.

Bloom, Nicholas (2009). "The Impact of Uncertainty Shocks." *Econometrica,* Vol. 77, pp. 623–85.

Blum, Avrim, John Hopcroft, and Ravindran Kannan (2016). *Foundations of Data Science.* Mimeo, Carnegie Mellon University.

Blume, Lawrence E., William A. Brock, Steven N. Durlauf, and Yannis M. Ioannides (2011). "Identification of Social Interactions." In *Handbook of Social Economics,* Vol. 1, pp. 853–964. Amsterdam: North-Holland.

Blumenstock, Joshua, and Xu Tan (2017). "Social Networks and Migration." Mimeo, University of Washington.

Blumer, Herbert (1958). "Race Prejudice as a Sense of Group Position." *Pacific Sociological Review,* Vol. 1:1, pp. 3–7.

Blundell, Richard, Luigi Pistaferri, and Ian Preston (2008). "Consumption Inequality and Partial Insurance." *The American Economic Review,* Vol. 98, pp. 1887–1921.

Bobo, Lawrence, Vincent L. Hutchings (1996). "Perceptions of Racial Group Competition: Extending Blumer's Theory of Group Position to a Multiracial Social Context." *American Sociological Review,* Vol. 61, pp. 951–72.

Bollobas, Bela (2001). *Random Graphs.* 2nd ed. Cambridge, U.K.: Cambridge University Press.

Boneva, Teodora, and Christopher Rauh (2015). "Parental Beliefs About Returns to Educational Investments—The Later the Better?" Available at SSRN: https://ssrn.com/abstract=2764288.

Borgatti, Stephen P. (2005). "Centrality and Network Flow." *Social Networks*, Vol. 27, pp. 55–71.

Borgatti, Stephen P., and Martin G. Everett (1992). "Notions of Position in Social Network Analysis." *Sociological Methodology*, pp. 1–35.

Borgatti, Stephen P., Martin G. Everett, and Jeffrey C. Johnson (2018). *Analyzing Social Networks*. Thousand Oaks, CA: Sage.

Borgatti, Stephen P., Candace Jones, and Martin G. Everett (1998). "Network Measures of Social Capital." *Connections*, Vol. 21, pp. 27–36.

Bourdieu, Pierre (1986). "The Forms of Capital" [In English]. In *Handbook of Theory and Research for the Sociology of Education*, pp. 241–58. Westport, CT: Greenwood Publishing Group.

Bourdieu, Pierre, and Jean Claude Passeron (1970). *La reproduction éléments pour une théorie du système d'enseignement*. Paris: Les Éditions de Minuit.

Bourguignon, François, and Christian Morrisson (2002). "Inequality Among World Citizens: 1820–1992." *The American Economic Review*, Vol. 92, pp. 727–44.

Boustan, Leah Platt (2016). *Competition in the Promised Land: Black Migrants in Northern Cities and Labor Markets*. Princeton, N.J.: Princeton University Press.

Bowles, Samuel, Eric Alden Smith, and Monique Borgerhoff Mulder, eds. (2010). "Inter-generational Wealth Transmission and Inequality in Premodern Societies." *Special Issue of Current Anthropology*, Vol. 51:1.

Boxell, Levi, Matthew Gentzkow, and Jesse M. Shapiro (2017). "Greater Internet Use Is Not Associated with Faster Growth in Political Polarization Among U.S. Demographic Groups." *Proceedings of the National Academy of Sciences*, Vol. 114 (40), pp. 10612–17.

Boyd, Robert, and Peter J. Richerson (1988). *Culture and the Evolutionary Process*. Chicago: University of Chicago Press.

Bradbury, Katharine (2011). "Trends in U.S. Family Income Mobility, 1969–2006." Federal Reserve Bank of Boston, Working Paper No. 11-10.

Bramoullé, Yann, and Brian W. Rogers (2010). "Diversity and Popularity in Social Networks." CIRPEE Working Paper No. 09-03.

Branch, Ben (2002). "The Costs of Bankruptcy: A Review." *International Review of Financial Analysis*, Vol. 11:1, pp. 39–57.

Brandts, Jordi, Ayça Ebru Giritligil, and Roberto A. Weber (2015). "An Experimental Study of Persuasion Bias and Social Influence in Networks." *European Economic Review*, Vol. 80, pp. 214–229.

Breen, Richard, and Jan O. Jonsson (2005). "Inequality of Opportunity in Comparative Perspective: Recent Research on Educational Attainment and Social Mobility." *Annual Review of Soccology*, Vol. 31, pp. 223–43.

Breiger, Ronald, and Pip Pattison (1986). "Cumulated Social Roles: The Duality of Persons and Their Algebras." *Social Networks*, Vol. 8, pp. 215–56.

Breza, Emily, and Arun G. Chandrasekhar (2016). "Social Networks, Reputation and Commitment: Evidence from a Savings Monitors Experiment." National Bureau of Economic Research Working Paper No. 21169.

Britton, Jack, Lorraine Dearden, Neil Shephard, and Anna Vignoles (2016). "How English Domiciled Graduate Earnings Vary with Gender, Institution Attended, Subject and Socio-Economic Background." Technical Report, Institute for Fiscal Studies.

Broido, Anna D., and Aaron Clauset (2018). "Scale-Free Networks Are Rare." arXiv preprint arXiv: 1801.03400.

Brown, Meta, Elizabeth Setren, and Giorgio Topa (2012). "Do Informal Referrals Lead to Better Matches? Evidence from a Firm's Employee Referral System." *FRB of New York Staff Report.*

Brummitt, Charles D., George Barnett, and Raissa M. D'Souza (2015). "Coupled Catastrophes: Sudden Shifts Cascade and Hop Among Interdependent Systems." *Journal of the Royal Society Interface,* Vol. 12:112, doi 10.1098/rsif.2015.0712.

Buchanan, James M., and Wm. Craig Stubblebine (1962). "Externality," *Economica,* Vol. 29:116, pp. 371–84.

Bullock, John G., Alan S. Gerber, Seth J. Hill, and Gregory A. Huber (2015). "Partisan Bias in Factual Beliefs about Politics." *Quarterly Journal of Political Science,* Vol. 10, pp. 519–78.

Bursztyn, Leonardo, Florian Ederer, Bruno Ferman, and Noam Yuchtman (2014). "Understanding Mechanisms Underlying Peer Effects: Evidence from a Field Experiment on Financial Decisions." *Econometrica,* Vol. 82, pp. 1273–1301.

Burt, Ronald S. (1992). *Structural Holes: The Social Structure of Competition.* Cambridge, Mass.: Harvard University Press.

——— (2000). "The Network Structure of Social Capital." *Research in Organizational Behavior,* Vol. 22, pp. 345–423.

——— (2005). *Brokerage and Closure: An Introduction to Social Capital.* Oxford, U.K.: Oxford University Press.

Cabrales, Antonio, Douglas Gale, and Piero Gottardi (2016). "Financial Contagion in Networks." In Bramoullé, Galeotti, and Rogers, eds., *The Oxford Handbook of the Economics of Networks.* Oxford, U.K.: Oxford University Press.

Cai, Jing, Alain de Janvry, and Elisabeth Sadoulet (2015). "Social Networks and the Decision to Insure." *American Economic Journal: Applied Economics,* Vol. 7:2, pp. 81–108.

Cai, Jing, and Adam Szeidl (2018). "Interfirm Relationships and Business Performance." *The Quarterly Journal of Economics.* Advance access: doi:10.1093/qje/qjx049.

Calvó-Armengol, Antoni, and Matthew O. Jackson (2004). "The Effects of Social Networks on Employment and Inequality." *The American Economic Review,* Vol. 94, pp. 426–54.

——(2007). "Networks in Labor Markets: Wage and Employment Dynamics and Inequality." *Journal of Economic Theory,* Vol. 132, pp. 27–46.

—— (2009). "Like Father, Like Son: Labor Market Networks and Social Mobility." *American Economic Journal: Microeconomics,* Vol. 1 (1), pp. 124–50.

Carayol, Nicolas, and Matthew O. Jackson (2017). "Evaluating the Underlying Qualities of Items and Raters from a Series of Reviews." Mimeo, Stanford University.

Card, David, and Laura Giuliano (2016). "Can Tracking Raise the Test Scores of High-Ability Minority Students?" *The American Economic Review,* Vol. 106, pp. 2783–2816.

Card, David, Alexandre Mas, and Jesse Rothstein (2008). "Tipping and the Dynamics of Segregation." *The Quarterly Journal of Economics,* Vol. 123 (1), pp. 177–218.

Carneiro, Pedro, and James J. Heckman (2002). "The Evidence on Credit Constraints in Post-Secondary Schooling." *The Economic Journal,* Vol. 112, pp. 705–34.

Carrell, Scott, and Bruce Sacerdote (2017). "Why Do College-Going Interventions Work?" *American Economic Journal: Applied Economics,* Vol. 9 (3), pp. 124–51.

Carrell, Scott E., Bruce I. Sacerdote, and James E. West (2013). "From Natural Variation to Optimal Policy? The Importance of Endogenous Peer Group Formation." *Econometrica,* Vol. 81, pp. 855–82.

Carrillo-Santisteve, P., and P. L. Lopalco (2012). "Measles Still Spreads in Europe: Who Is Responsible for the Failure to Vaccinate?" *Clinical Microbiology and Infection,* Vol. 18, pp. 50–56.

Carrington, William J., Enrica Detragiache, and Tara Vishwanath (1996). "Migration with Endogenous Moving Costs." *The American Economic Review,* Vol. 86, pp. 909–30.

Carter, Michael R., and Christopher B. Barrett (2006). "The Economics of Poverty Traps and Persistent Poverty: An Asset-Based Approach." *The Journal of Development Studies,* Vol. 42, pp. 178–99.

Carvalho, Vasco M. (2014). "From Micro to Macro Via Production Networks." *Journal of Economic Perspectives,* Vol. 28 (4), pp. 23–48.

Caulier, Jean-François, Ana Mauleon, and Vincent Vannetelbosch (2013). "Contractually Stable Networks." *International Journal of Game Theory,* Vol. 42 (2), pp. 483–99.

Centola, D. (2011). "An Experimental Study of Homophily in the Adoption of Health Behavior." *Science,* Vol. 334:6060, pp. 1269–72.

Centola, Damon, Víctor M. Eguíluz, and Michael W. Macy (2007). "Cascade Dynamics of Complex Propagation." *Physica A: Statistical Mechanics and Its Applications,* Vol. 374, pp. 449–56.

Cépon, Bruno, Florencia Devoto, Esther Duflo, and William Parienté (2015). "Estimating the Impact of Microcredit on Those Who Take It

Up: Evidence from a Randomized Experiment in Morocco." *American Economic Journal: Applied Economics*, Vol. 7, pp. 123–50.

Cesaretti, Rudolf, José Lobo, Luís M.A. Bettencourt, Scott G. Ortman, and Michael E. Smith (2016). "Population-Area Relationship for Medieval European Cities." *PLoS ONE*, Vol. 11 (10), article no. e0162678.

Chandrasekhar, Arun, and Matthew O. Jackson (2013). "Tractable and Consistent Random Graph Models." Available at SSRN: http://ssrn.com /abstract=2150428.

———— (2016). "A Network Formation Model Based on Subgraphs." Available at SSRN: https://ssrn.com/abstract=2660381.

Chandrasekhar, Arun G., Horacio Larreguy, and Juan Pablo Xandri (2015). "Testing Models of Social Learning on Networks: Evidence from a Lab Experiment in the Field." National Bureau of Economic Research Paper No. 21468.

Chaney, Thomas (2014). "The Network Structure of International Trade." *The American Economic Review*, Vol. 104, pp. 3600–3634.

Chen, Wei, Zhenming Liu, Xiaorui Sun, and Yajun Wang (2010). "A Game-Theoretic Framework to Identify Overlapping Communities in Social Networks." *Data Mining and Knowledge Discovery*, Vol. 21, pp. 224–40.

Chen, Yan, and Sherry Xin Li (2009). "Group Identity and Social Preferences." *The American Economic Review*, Vol. 99, pp. 431–57.

Chetty, Raj, John N. Friedman, Emmanuel Saez, Nicholas Turner, and Danny Yagan (2017). "Mobility Report Cards: The Role of Colleges in Intergenerational Mobility," National Bureau of Economic Research Working Paper No. 23618.

Chetty, Raj, David Grusky, Maximilian Hell, Nathaniel Hendren, Robert Manduca, and Jimmy Narang (2017). "The Fading American Dream: Trends in Absolute Income Mobility Since 1940." *Science*, Vol. 356 (6336), pp. 398–406.

Chetty, Raj, and Nathaniel Hendren (2015). "The Impacts of Neighborhoods on Intergenerational Mobility: Childhood Exposure Effects and County-Level Estimates." Harvard University and National Bureau of Economic Research Working Paper No. 23001.

Chetty, Raj, Nathaniel Hendren, and Lawrence F. Katz (2016b). "The Effects of Exposure to Better Neighborhoods on Children: New Evidence from the Moving to Opportunity Experiment." *The American Economic Review*, Vol. 106, pp. 855–902.

Chiappori, Pierre-André, Bernard Salanié, and Yoram Weiss (2017). "Partner Choice, Investment in Children, and the Marital College Premium." *The American Economic Review*, Vol. 107 (8), pp. 2109–67.

Choi, S., D. Gale, and S. Kariv (2005). "Behavioral Aspects of Learning in Social Networks: An Experimental Study." *Advances in Applied Microeconomics: A Research Annual*, Vol. 13, pp. 25–61.

Christakis, Nicholas A., and James H. Fowler (2009). *Connected: The Surprising Power of Our Social Networks and How They Shape Our Lives.* New York: Little, Brown.

—— (2010). "Social Network Sensors for Early Detection of Contagious Outbreaks." *PLoS ONE,* Vol. 5 (9), paper e12948.

—— (2014). "Friendship and Natural Selection." *Proceedings of the National Academy of Sciences,* Vol. 111, pp. 10796–801.

Clampet-Lundquist, Susan, and Douglas S. Massey (2008). "Neighborhood Effects on Economic Self-Sufficiency: A Reconsideration of the Moving to Opportunity Experiment." *American Journal of Sociology,* Vol. 114, pp. 107–143.

Clauset, Aaron (2018). "Trends and Fluctuations in the Severity of Interstate Wars." *Science Advances,* Vol. 4 (2), article no. eaao3580.

Clauset, Aaron, Mark E. J. Newman, and Cristopher Moore (2004). "Finding Community Structure in Very Large Networks." *Physical Review E,* Vol. 70, article no. 066111.

Clauset, Aaron, Cosma R. Shalizi, and Mark E. J. Newman (2009). "Power-Law Distributions in Empirical Data." *SIAM Review,* Vol. 51, pp. 661–703.

Clifford, Peter, and Aidan Sudbury (1973). "A Model for Spatial Conflict." *Biometrika,* Vol. 60, pp. 581–88.

Coase, Ronald H. (1960). "The Problem of Social Cost." *Journal of Law and Economics,* Vol. 3, pp. 1–44.

Coffman, Lucas C., Clayton R. Featherstone, and Judd B. Kessler (2016). "Can Social Information Affect What Job You Choose and Keep?" *American Economic Journal: Applied Economics,* Vol. 9 (1), pp. 96–117.

Coleman, James S. (1961). *The Adolescent Society.* New York: Free Press.

—— (1988). "Social Capital in the Creation of Human Capital." *American Journal of Sociology,* Vol. 94, pp. S95–S120.

Coleman, James S., and Thomas Hoffer (1987). *Public and Private High Schools: The Impact of Communities.* New York: Basic Books.

Coman, Alin, Ida Momennejad, Rae D. Drach, and Andra Geana (2016). "Mnemonic Convergence in Social Networks: The Emergent Properties of Cognition at a Collective Level." *Proceedings of the National Academy of Sciences,* 113 (29), pp. 8171–76.

Conley, Timothy G., and Christopher R. Udry (2010). "Learning About a New Technology: Pineapple in Ghana." *The American Economic Review,* Vol. 100, pp. 35–69.

Copic, Jernej, Matthew O. Jackson, and Alan Kirman (2009). "Identifying Community Structures from Network Data via Maximum Likelihood Methods." *The BE Journal of Theoretical Economics,* Vol. 9 (1), article no. 1935-1704.

Corak, Miles (2016). "Inequality from Generation to Generation: The United States in Comparison." IZA Discussion Paper No. 9929.

Cowling, Benjamin J., Lincoln L. H. Lau, Peng Wu, Helen W. C. Wong, Vicky J. Fang, Steven Riley, and Hiroshi Nishiura (2010). "Entry Screening to Delay Local Transmission of 2009 Pandemic Influenza A (H1N1)." *BMC Infectious Diseases*, Vol. 10, pp. 1–4.

Cunha, Flavio (2016). "Gaps in Early Investments in Children." Preprint, Rice University.

Cunha, Flavio, James J. Heckman, Lance Lochner, and Dimitriy V. Masterov (2006). "Interpreting the Evidence on Life Cycle Skill Formation." *Handbook of the Economics of Education*, Vol. 1, pp. 697–812.

Currarini, Sergio, Matthew O. Jackson, and Paolo Pin (2009). "An Economic Model of Friendship: Homophily, Minorities, and Segregation." *Econometrica*, Vol. 77, pp. 1003–45.

——— (2010). "Identifying the Roles of Race-Based Choice and Chance in High School Friendship Network Formation." *Proceedings of the National Academy of Sciences*, Vol. 107, pp. 4857–61.

Currarini, Sergio, and Massimo Morelli (2000). "Network Formation with Sequential Demands." *Review of Economic Design*, Vol. 5, pp. 229–50.

Currie, Janet (2001). "Early Childhood Education Programs." *Journal of Economic Perspectives*, Vol. 15 (2), pp. 213–38.

Cushing, Frank H. (1981). *Zuñi: Selected Writings of Frank Hamilton Cushing*. Edited, with an introduction by Jesse Green. Lincoln: University of Nebraska Press.

Cutler, David M., and Edward L Glaeser (2010). "Social Interactions and Smoking." *Research Findings in the Economics of Aging*, pp. 123–41.

Dasgupta, Partha (2005). "Economics of Social Capital." *Economic Record*, Vol. 81 (s1), pp. 1–21.

Dasgupta, Partha, and Ismail Serageldin (2001). *Social Capital: A Multifaceted Perspective*. World Bank Publications.

Davidow, William H. (2011). *Overconnected: The Promise and Threat of the Internet*. San Francisco: Open Road Media.

Davis, James H. (1992). "Some Compelling Intuitions About Group Consensus Decisions, Theoretical and Empirical Research, and Interpersonal Aggregation Phenomena: Selected Examples 1950–1990." *Organizational Behavior and Human Decision Processes*, Vol. 52, pp. 3–38.

Davis, Nicholas T., and Johanna L. Dunaway (2016). "Party Polarization, Media Choice, and Mass Partisan-Ideological Sorting." *Public Opinion Quarterly*, Vol. 80 (s1), pp. 272–97.

Dawkins, Richard (1976). *The Selfish Gene*. Oxford, U.K.: Oxford University Press.

Dean, Katharine R., Fabienne Krauer, Lars Walløe, Ole Christian Lingjærde, Barbara Bramanti, Nils Chr Stenseth, and Boris V. Schmid (2018). "Human Ectoparasites and the Spread of Plague in Europe During the Second Pandemic." *Proceedings of the National Academy of Sciences*, Vol. 115 (6), pp. 1304–09.

DeJong, William, Shari K. Schneider, Laura G. Towvim, Melissa J. Murphy, Emily E. Doerr, Neil R. Simonsen, Karen E. Mason, and Richard A. Scribner (2006). "A Multisite Randomized Trial of Social Norms Marketing Campaigns to Reduce College Student Drinking." *Journal of Studies on Alcohol*, Vol. 67, pp. 868–79.

DeMarzo, Peter M., Dimitri Vayanos, and Jeffrey Zwiebel (2003). "Persuasion Bias, Social Influence, and Unidimensional Opinions." *The Quarterly Journal of Economics*, Vol. 118 (3), pp. 909–968.

Demirer, Mert, Francis X. Diebold, Laura Liu, and Kamil Yilmaz (2018). "Estimating Global Bank Network Connectedness." *Journal of Applied Econometrics*, Vol. 33 (1), pp. 1–15.

de Souza Briggs, Xavier, Susan J. Popkin, and John Goering (2010). *Moving to Opportunity: The Story of an American Experiment to Fight Ghetto Poverty*. Oxford, UK: Oxford University Press.

Dhillon, Amrita, Vegard Iversen, and Gaute Torsvik (2013). "Employee Referral, Social Proximity, and Worker Discipline." Mimeo, University of Warwick.

Diamond, Jared (1997). *Guns, Germs, and Steel*. New York: W. W. Norton.

Diestel, R. (2000). "Graph Theory." *Graduate Texts in Mathematics*, 173. Berlin and Heidelberg: Springer-Verlag.

Dodds, Peter S., Roby Muhamad, and Duncan J. Watts (2003). "An Experimental Study of Search in Global Social Networks." *Science*, Vol. 301, pp. 827–29.

Doleac, Jennifer L., and Benjamin Hansen (2016). "Does Ban the Box Help or Hurt Low-Skilled Workers? Statistical Discrimination and Employment Outcomes When Criminal Histories Are Hidden." National Bureau of Economic Research Working Paper No. 22469.

Domingue, Benjamin W., Daniel W. Belsky, Jason M. Fletcher, Dalton Conley, Jason D. Boardman, and Kathleen Mullan Harris (2018). "The Social Genome of Friends and Schoolmates in the National Longitudinal Study of Adolescent to Adult Health." *Proceedings of the National Academy of Sciences*, Vol 115 (4), pp. 702–07.

Duan, Wenjing, Bin Gu, and Andrew B. Whinston (2008). "Do Online Reviews Matter?—An Empirical Investigation of Panel Data." *Decision Support Systems*, Vol. 45 (4), pp. 1007–16.

Dubois, Florent, and Christophe Muller (2016). "Segregation and the Perception of the Minority." Mimeo, University of Aix-Marseille.

Duboscq, Julie, Valeria Romano, Cédric Sueur, and Andrew J. J. MacIntosh (2016). "Network Centrality and Seasonality Interact to Predict Lice Load in a Social Primate." *Scientific Reports*, Vol. 6, article no. 22095.

Duflo, Esther, and Emanuel Saez (2003). "The Role of Information and Social Interactions in Retirement Plan Decisions: Evidence from a Randomized Experiment." *The Quarterly Journal of Economics*, Vol. 118 (3), pp. 815–42.

Dupont, Brandon (2007). "Bank Runs, Information and Contagion in the Panic of 1893." *Explorations in Economic History,* Vol. 44, pp. 411–31.

Durlauf, Steven N. (1996). "A Theory of Persistent Income Inequality." *Journal of Economic Growth,* Vol. 1, pp. 75–93.

Dutta, Bhaskar, and Matthew O. Jackson (2000). "The Stability and Efficiency of Directed Communication Networks." *Review of Economic Design,* Vol. 5, pp. 251–72.

Dutta, Bhaskar, and Suresh Mutuswami (1997). "Stable Networks." *Journal of Economic Theory,* Vol. 76, pp. 322–44.

Easley, David, and Jon Kleinberg (2010). *Networks, Crowds, and Markets: Reasoning About a Highly Connected World.* Cambridge, U.K.: Cambridge University Press.

Easterly, William (2009). "Empirics of Strategic Interdependence: The Case of the Racial Tipping Point," *The B.E. Journal of Macroeconomics,* Vol. 9 (1), article no. 25.

Eberhardt, Jennifer L., Phillip Atiba Goff, Valerie J. Purdie, and Paul G. Davies (2004). "Seeing Black: Race, Crime, and Visual Processing," *Journal of Personality and Social Psychology,* Vol. 87, p. 876.

Echenique, Federico, and Roland G. Fryer, Jr. (2007). "A Measure of Segregation Based on Social Interactions." *The Quarterly Journal of Economics,* Vol. 122 (2), pp. 441–85.

Eckel, Catherine C., and Philip J. Grossman (2005). "Managing Diversity by Creating Team Identity." *Journal of Economic Behavior & Organization,* Vol. 58, pp. 371–92.

Edling, Christofer, Jens Rydgren, and Rickard Sandell (2016). "Terrorism, Belief Formation, and Residential Integration: Population Dynamics in the Aftermath of the 2004 Madrid Terror Bombings." *American Behavioral Scientist,* Vol. 60 (10), pp. 1215–31.

Elliott, Matthew, Benjamin Golub, and Matthew O. Jackson (2014). "Financial Networks and Contagion." *The American Economic Review,* Vol. 104 (10), pp. 3115–53.

England, Paula (2017). *Comparable Worth: Theories and Evidence.* New York: Routledge.

Enke, Benjamin, and Florian Zimmermann (2017). "Correlation Neglect in Belief Formation." *Review of Economic Studies,* forthcoming.

Eom, Young-Ho, and Hang-Hyun Jo (2014). "Generalized Friendship Paradox in Complex Networks: The Case of Scientific Collaboration." *Scientific Reports,* Vol. 4, article no. 4603.

Epple, Dennis, and Richard Romano (2011). "Peer Effects in Education: A Survey of the Theory and Evidence." In *The Handbook of Social Economics,* Vol. 1., pp. 1053–1163, ed. Jess Benhabib, Alberto Bisin, and Matthew O. Jackson. Amsterdam: North-Holland.

Ercsey-Ravasz, Mária, Ryan N. Lichtenwalter, Nitesh V. Chawla, and Zoltán

Toroczkai (2012). "Range-Limited Centrality Measures in Complex Networks." *Physical Review E*, Vol. 85, p. 066103.

Erdős, Paul, and Alfréd Rényi (1959). "On Random Graphs." *Publicationes Mathematicae Debrecen*, Vol. 6, p. 156.

—— (1960). "On the Evolution of Random Graphs." *Publ. Math. Inst. Hung. Acad. Sci*, Vol. 5, pp. 17–60.

Everett, Martin G., and Stephen P. Borgatti (1999). "The Centrality of Groups and Classes." *The Journal of Mathematical Sociology*, Vol. 23, pp. 181–201.

—— (2005). "Ego Network Betweenness." *Social Networks*, Vol. 27, pp. 31–38.

Everett, Martin G., and Thomas W. Valente (2016). "Bridging, Brokerage and Betweenness." *Social Networks*, Vol. 44, pp. 202–8.

Fafchamps, Marcel, and Flore Gubert (2007). "The Formation of Risk Sharing Networks." *Journal of Development Economics*, Vol. 83, pp. 326–50.

Fafchamps, Marcel, and Susan Lund (2003). "Risk-Sharing Networks in Rural Philippines." *Journal of Development Economics*, Vol. 71, pp. 261–87.

Fafchamps, Marcel, Marco J. van der Leij, and Sanjeev Goyal (2010). "Matching and Network Effects." *Journal of the European Economic Association*, Vol. 8, pp. 203–31.

Farboodi, Maryam (2014). "Intermediation and Voluntary Exposure to Counterparty Risk," Mimeo, Princeton University.

Feigenberg, Benjamin, Erica Field, and Rohini Pande (2013). "The Economic Returns to Social Interaction: Experimental Evidence from Microfinance." *The Review of Economic Studies*, Vol. 80 (4), pp. 1459–83.

Feinberg, Matthew, Robb Willer, and Michael Schultz (2014). "Gossip and Ostracism Promote Cooperation in Groups." *Psychological Science*, Vol. 25 (3), pp. 656–64.

Feld, Scott L. (1991). "Why Your Friends Have More Friends than You Do." *American Journal of Sociology*, Vol. 96 (6), pp. 1464–77.

Felfe, Christina, and Rafael Lalive (2018). "Does Early Childcare Affect Children's Development." Mimeo, University of Lausanne.

Ferguson, Neil M., Derek A. T. Cummings, Christophe Fraser, James C. Cajka, Philip C. Cooley, and Donald S. Burke (2006). "Strategies for Mitigating an Influenza Pandemic." *Nature*, Vol. 442, pp. 448–52.

Ferguson, Niall (2018). *The Square and the Tower: Networks, Hierarchies and the Struggle for Global Power*. London: Penguin.

Fernandez, Marilyn, and Laura Nichols (2002). "Bridging and Bonding Capital: Pluralist Ethnic Relations in Silicon Valley." *International Journal of Sociology and Social Policy*, Vol. 22, pp. 104–22.

Fernandez, Roberto M., Emilio J. Castilla, and Paul Moore (2000). "Social Capital at Work: Networks and Employment at a Phone Center." *American Journal of Sociology*, pp. 1288–1356.

Fichtner, Paula Sutter (1976). "Dynastic Marriage in Sixteenth-Century Habsburg Diplomacy and Statecraft: An Interdisciplinary Approach." *The American Historical Review*, Vol. 81, pp. 243–65.

Field, Erica, and Rohini Pande (2008). "Repayment Frequency and Default in Microfinance: Evidence from India." *Journal of the European Economic Association*, Vol. 6, pp. 501–9.

Fiorina, Morris P. (2017). "The Political Parties Have Sorted." Hoover Institution Essay on Contemporary Politics, Series No. 3.

Fisher, Len (2009). *The Perfect Swarm: The Science of Complexity in Everyday Life*. New York: Basic Books.

Fisman, Raymond, and Edward Miguel (2007). "Corruption, Norms, and Legal Enforcement: Evidence from Diplomatic Parking Tickets." *Journal of Political Economy*, Vol. 115, pp. 1020–48.

Fisman, Raymond, Jing Shi, Yongxiang Wang, and Rong Xu (2018). "Social Ties and Favoritism in Chinese Science." *Journal of Political Economy*, forthcoming.

Fleurbaey, Marc, and François Maniquet (2011). *A Theory of Fairness and Social Welfare*. Cambridge, U.K.: Cambridge University Press.

Fortunato, Santo (2010). "Community Detection in Graphs." *Physics Reports*, Vol. 486 (3–5), pp. 75–174.

Foster, Andrew D., and Mark R. Rosenzweig (1995). "Learning by Doing and Learning from Others: Human Capital and Technical Change in Agriculture." *Journal of Political Economy*, Vol. 103, pp. 1176–1209.

Fowler, James H., Christopher T. Dawes, and Nicholas A. Christakis (2009). "Model of Genetic Variation in Human Social Networks." *Proceedings of the National Academy of Sciences*, Vol. 106, pp. 1720–24.

Fradkin, Andrey, Elena Grewal, Dave Holtz, and Matthew Pearson (2015). "Bias and Reciprocity in Online Reviews: Evidence from Field Experiments on Airbnb." In *Proceedings of the Sixteenth ACM Conference on Economics and Computation*, pp. 641–641.

Frederick, Shane (2012). "Overestimating Others' Willingness to Pay." *Journal of Consumer Research*, Vol. 39 (1), pp. 1–21.

Freeman, Linton C. (1977). "A Set of Measures of Centrality Based on Betweenness." *Sociometry*, Vol. 40 (1), pp. 35–41.

—— (1978). "Centrality in Social Networks Conceptual Clarification." *Social Networks*, Vol. 1, pp. 215–39.

French, J. R. (1956). "A Formal Theory of Social Power." *Psychological Review*, Vol. 63 (3), pp. 181–94.

Fricke, Daniel, and Thomas Lux (2014). "Core Periphery Structure in the Overnight Money Market: Evidence from the e-MID Trading Platform." *Computational Economics*, Vol. 45 (3), pp. 359–95.

Friedman, Thomas L. (2016). *Thank You for Being Late: An Optimist's Guide to Thriving in the Age of Accelerations*. New York: Farrar, Straus & Giroux.

Froot, Kenneth A., David S. Scharfstein, and Jeremy C. Stein (1992). "Herd on the Street: Informational Inefficiencies in a Market with Short-Term Speculation." *The Journal of Finance*, Vol. 47, pp. 1461–84.

Fryer, Roland G., Jr. (2007). "Guess Who's Been Coming to Dinner? Trends in Interracial Marriage over the 20th Century." *Journal of Economic Perspectives*, Vol. 21 (2), pp. 71–90.

Fryer, Roland G., Jr., Jacob K. Goeree, and Charles A. Holt (2005). "Experience-Based Discrimination: Classroom Games." *The Journal of Economic Education*, Vol. 36, pp. 160–70.

Fryer, Roland G., Jr., and Lawrence F. Katz (2013). "Achieving Escape Velocity: Neighborhood and School Interventions to Reduce Persistent Inequality." *The American Economic Review*, Vol. 103, pp. 232–37.

Fryer, Roland G., Jr., and Steven D. Levitt (2004). "The Causes and Consequences of Distinctively Black Names." *The Quarterly Journal of Economics*, pp. 767–805.

Fryer, Roland G., Jr., Steven D. Levitt, and John A. List (2015). "Parental Incentives and Early Childhood Achievement: A Field Experiment in Chicago Heights." National Bureau of Economic Research Working Paper No. 21477.

Gächter, Simon, and Jonathan F. Schulz (2016). "Intrinsic Honesty and the Prevalence of Rule Violations Across Societies." *Nature*, Vol. 531, pp. 496–99.

Gaertner, Samuel L., John F. Dovidio, Phyllis A. Anastasio, Betty A. Bachman, and Mary C. Rust (1993). "The Common Ingroup Identity Model: Recategorization and the Reduction of Intergroup Bias." *European Review of Social Psychology*, Vol. 4, pp. 1–26.

Gagnon, Julien, and Sanjeev Goyal (2017). "Networks, Markets, and Inequality." *The American Economic Review*, Vol. 107, pp. 1–30.

Gai, Prasanna, and Sujit Kapadia (2010). "Contagion in Financial Networks." *Proceedings of the Royal Society A*, Vol. 466, pp. 2401–23.

Galbiati, Roberto, and Giulio Zanella (2012). "The Tax Evasion Social Multiplier: Evidence from Italy." *Journal of Public Economics*, Vol. 96, pp. 485–94.

Galenianos, Manolis (2016). "Referral Networks and Inequality." SSRN Paper No. 2768083.

Galeotti, Andrea, and Sanjeev Goyal (2010). "The Law of the Few." *The American Economic Review*, Vol. 100 (4), pp. 1468–92.

Galeotti, Andrea, Sanjeev Goyal, Matthew O. Jackson, Fernando Vega-Redondo, and Leeat Yariv (2010). "Network Games." *The Review of Economic Studies*, Vol. 77 (1), pp. 218–44.

Galton, Francis (1901). "Vox Populi." *Nature*, Vol. 75 (7), pp. 450–51.

Garas, Antonios, Panos Argyrakis, Céline Rozenblat, Marco Tomassini, and Shlomo Havlin (2010). "Worldwide Spreading of Economic Crisis." *New Journal of Physics*, Vol. 12, article no. 113043.

Garces, Eliana, Duncan Thomas, and Janet Currie (2002). "Longer-Term Effects of Head Start." *The American Economic Review*, Vol. 92, pp. 999–1012.

Gee, Laura K., Jason Jones, and Moira Burke (2017). "Social Networks and Labor Markets: How Strong Ties Relate to Job Finding on Facebook's Social Network." *Journal of Labor Economics*, Vol. 35, pp. 485–518.

Gentzkow, Matthew (2017). "Polarization in 2016." Essay, Stanford University.

Gentzkow, Matthew, Jesse M. Shapiro, and Matt Taddy (2016). "Measuring Polarization in High-Dimensional Data: Method and Application to Congressional Speech." National Bureau of Economic Research Working Paper No. 22423.

Gerber, Jeffrey S., and Paul A. Offit (2009). "Vaccines and Autism: A Tale of Shifting Hypotheses." *Clinical Infectious Diseases*, Vol. 48, pp. 456–61.

Ghiglino, Christian, and Sanjeev Goyal (2010). "Keeping Up with the Neighbors: Social Interaction in a Market Economy." *Journal of the European Economic Association*, Vol. 8 (1), pp. 90–119.

Giancola, Jennifer, and Richard D. Kahlenberg (2016). "True Merit: Ensuring Our Brightest Students Have Access to Our Best Colleges and Universities." Jack Kent Cooke Foundation.

Gilbert, Edgar N. (1959). "Random Graphs." *The Annals of Mathematical Statistics*, Vol. 30, pp. 1141–44.

Gilchrist, Duncan Sheppard, and Emily Glassberg Sands (2016). "Something to Talk About: Social Spillovers in Movie Consumption." *Journal of Political Economy*, Vol. 124, pp. 1339–82.

Gilman, Eric, Shelley Clarke, Nigel Brothers, Joanna Alfaro-Shigueto, John Mandelman, Jeff Mangel, Samantha Petersen, Susanna Piovano, Nicola Thomson, Paul Dalzell, Miguel Donosol, Meidad Gorenh, and Tim Werner (2008). "Shark Interactions in Pelagic Longline Fisheries." *Marine Policy*, Vol. 32, pp. 1–18.

Gjerstad, Steven (2004). "Risk Aversion, Beliefs, and Prediction Market Equilibrium." Economic Science Laboratory Working Paper 04-17, University of Arizona.

Gladwell, Malcolm (2000). *The Tipping Point: How Little Things Can Make a Big Difference*. New York: Back Bay Books.

Glaeser, Edward L., Bruce I. Sacerdote, and Jose A. Scheinkman (2003). "The Social Multiplier." *Journal of the European Economic Association*, Vol. 1, pp. 345–53.

Glaeser, Edward L., and Cass R. Sunstein (2009). "Extremism and Social Learning." *Journal of Legal Analysis*, Vol. 1, pp. 263–324.

Glasserman, Paul, and H. Peyton Young (2016). "Contagion in Financial Networks." *Journal of Economic Literature*, Vol. 54 (3), pp. 779–831.

Gneezy, Uri, John List, and Michael K. Price (2012). "Toward an Understanding of Why People Discriminate: Evidence from a Series of Natural Field Experiments." National Bureau of Economic Research Working Paper No.17855.

Godfrey, Stephanie S. (2013). "Networks and the Ecology of Parasite Transmission: A Framework for Wildlife Parasitology." *International Journal for Parasitology: Parasites and Wildlife*, Vol. 2, pp. 235–45.

Godfrey, Stephanie S., Jennifer A. Moore, Nicola J. Nelson, and C. Michael

Bull (2010). "Social Network Structure and Parasite Infection Patterns in a Territorial Reptile, the Tuatara (Sphenodon punctatus)." *International Journal for Parasitology*, Vol. 40 (13), pp. 1575–85.

Goel, Rahul, Sandeep Soni, Naman Goyal, John Paparrizos, Hanna Wallach, Fernando Diaz, and Jacob Eisenstein (2016). "The Social Dynamics of Language Change in Online Networks." In *International Conference on Social Informatics*, pp. 41–57. New York: Springer.

Goel, Sharad, Ashton Anderson, Jake Hofman, and Duncan J. Watts (2015). "The Structural Virality of Online Diffusion." *Management Science*, Vol. 62 (1), pp. 180–96.

Gofman, Michael (2011). "A Network-Based Analysis of Over-the-Counter Markets." Available at SSRN: https://dx.doi.org/10.2139/ssrn.1681151.

Goldin, Claudia D., and Lawrence F. Katz (2009). *The Race Between Education and Technology*. Cambridge, Mass.: Harvard University Press.

Golub, Benjamin, and Matthew O. Jackson (2010). "Naive Learning in Social Networks and the Wisdom of Crowds." *American Economic Journal: Microeconomics*, Vol. 2, pp. 112–49.

—— (2012). "How Homophily Affects the Speed of Learning and Best-Response Dynamics." *The Quarterly Journal of Economics*, Vol. 127, pp. 1287–1338.

Gould, Roger V., and Roberto M. Fernandez (1989). "Structures of Mediation: A Formal Approach to Brokerage in Transaction Networks." *Sociological Methodology*, Vol. 19, pp. 89–126.

Goyal, S. (2007). *Connections*. Princeton, N.J.: Princeton University Press.

Granovetter, Mark S. (1973). "The Strength of Weak Ties." *The American Journal of Sociology*, Vol. 78, pp. 1360–80.

—— (1978). "Threshold Models of Collective Behavior." *The American Journal of Sociology*, Vol. 83, pp. 1420–43.

—— (1985). "Economic Action and Social Structure: The Problem of Embeddedness." *The American Journal of Sociology*, Vol. 91, pp. 481–510.

—— (1995). *Getting a Job: A Study of Contacts and Careers*. 2nd ed. Chicago: University of Chicago Press.

Grinblatt, Mark, Matti Keloharju, and Seppo Ikäheimo (2008). "Social Influence and Consumption: Evidence from the Automobile Purchases of Neighbors." *The Review of Economics and Statistics*, Vol. 90, pp. 735–53.

Grusky, David B., and Manwai C. Ku (2008). "Gloom, Doom, and Inequality." In *Social Stratification: Class, Race, and Gender in Sociological Perspective*, Vol. 3, pp. 2–28.

Haden, Brian (1995). "Chapter 2: Pathways to Power: Principles for Creating Socioeconomic Inequalities." In T. Douglas Price and Gary M. Feinman, eds., *Fundamental Issues in Archaeology*, pp. 15–86. New York: Springer.

Hahn, Robert W., and Paul C. Tetlock (2006). *Information Markets: A New Way of Making Decisions*. New York: AEI-Brookings Joint Center for Regulatory Studies.

Haines, Michael, and S. F. Spear (1996). "Changing the Perception of the Norm: A Strategy to Decrease Binge Drinking Among College Students." *Journal of American College Health*, Vol. 45, pp. 134–40.

Hamilton James T. (2016). *Democracy's Detectives: The Economics of Investigative Journalism*. Cambridge, Mass.: Harvard University Press.

Hampton, Keith, and Barry Wellman (2003). "Neighboring in Netville: How the Internet Supports Community and Social Capital in a Wired Suburb." *City & Community*, Vol. 2, pp. 277–311.

Haney, Craig, Curtis Banks, and Philip Zimbardo (1973). "Interpersonal Dynamics in a Simulated Prison." *International Journal of Criminology and Penology*, Vol. 1, pp. 69–97.

Hanifan, Lyda J. (1916). "The Rural School Community Center." *The Annals of the American Academy of Political and Social Science*, Vol. 67, pp. 130–38.

——— (1920). *The Community Center*. London: Silver, Burdett.

Harary, Frank (1959). "A Criterion for Unanimity in French's Theory of Social Power," in D. Cartwright, ed., *Studies in Social Power*. Ann Arbor: University of Michigan Press.

Hart, Betty, and Todd R. Risley (1995). *Meaningful Differences in the Everyday Experience of Young American Children*. Baltimore: Paul H. Brookes Publishing.

Hayden, Brian (1997). *The Pithouses of Keatley Creek*. New York: Harcourt Brace College.

Heath, Rachel (forthcoming). "Why Do Firms Hire Using Referrals? Evidence from Bangladeshi Garment Factories." *Journal of Political Economy*.

Heckman, James J. (2012). "Invest in Early Childhood Development: Reduce Deficits, Strengthen the Economy." *The Heckman Equation*, Vol. 7, pp. 1–2.

Heckman, James J., Seong Hyeok Moon, Rodrigo Pinto, Peter A. Savelyev, and Adam Yavitz (2010). "The Rate of Return to the HighScope Perry Preschool Program." *Journal of Public Economics*, Vol. 94, pp. 114–28.

Henrich, Joseph (2015). *The Secret of Our Success: How Culture Is Driving Human Evolution, Domesticating Our Species, and Making Us Smarter*. Princeton, N.J.: Princeton University Press.

Henrich, Joseph, Robert Boyd, Samuel Bowles, Colin Camerer, Ernst Fehr, Herbert Gintis, and Richard McElreath (2001). "In Search of Homo Economicus: Behavioral Experiments in 15 Small-Scale Societies." *The American Economic Review*, Vol. 91, pp. 73–78.

Herings, P. Jean-Jacques, Ana Mauleon, and Vincent Vannetelbosch (2009). "Farsightedly Stable Networks." *Games and Economic Behavior*, Vol. 67, pp. 526–41.

Hicks, Michael J., and Srikant Devaraj (2015). "The Myth and the Reality of Manufacturing in America." Center for Business and Economic Research, Ball State University.

Hilger, Nathaniel (2016). "Upward Mobility and Discrimination: The Case of Asian Americans." National Bureau of Economic Research Working Paper No. 22748.

Hill, Kim (2002). "Altruistic Cooperation During Foraging by the Ache, and the Evolved Human Predisposition to Cooperate." *Human Nature*, Vol. 13, pp. 105–28.

Hill, Kim, and A. Magdalena Hurtado (2017). *Ache Life History: The Ecology and Demography of a Foraging People*. New York: Routledge.

Hillebrand, Evan (2009). "Poverty, Growth, and Inequality over the Next 50 Years." Expert Meeting on How to Feed the World in 2050, Food and Agriculture Organization of the United Nations Economic and Social Development Department.

Hinsz, Verlin B., R. Scott Tindale, and David A. Vollrath (1997). "The Emerging Conceptualization of Groups as Information Processors." *Psychological Bulletin*, Vol. 121, p. 43.

Hirshleifer, David, and Siew Hong Teoh (2003). "Herd Behaviour and Cascading in Capital Markets: A Review and Synthesis." *European Financial Management*, Vol. 9, pp. 25–66.

Hjort, Jonas, and Jonas Poulsen (2018). "The Arrival of Fast Internet and Employment in Africa." National Bureau of Economic Research Working Paper No. 23582.

Hodas, Nathan O., Farshad Kooti, and Kristina Lerman (2013). "Friendship Paradox Redux: Your Friends Are More Interesting than You." ArXiv: 1304.3480v1.

Hofstra, Bas, Rense Corten, Frank van Tubergen, and Nicole B. Ellisond (2017). "Sources of Segregation in Social Networks: A Novel Approach Using Facebook." *American Sociological Review*, Vol. 82 (3), pp. 625–56.

Hogg, Michael A., and Scott Tindale (2008). *Blackwell Handbook of Social Psychology: Group Processes*. New York: John Wiley & Sons.

Holland, Paul W., Kathryn Blackmond Laskey, and Samuel Leinhardt (1983). "Stochastic Blockmodels: First Steps." *Social Networks*, Vol. 5, pp. 109–37.

Hölldbler, Bert, and Edward O. Wilson (1990). *The Ants*. Cambridge, Mass.: Harvard University Press.

Holley, Richard A., and Thomas M. Liggett (1975). "Ergodic Theorems for Weakly Interacting Infinite Systems and the Voter Model." *The Annals of Probability*, Vol. 3 (4), pp. 643–63.

Hoxby, Caroline, and Christopher Avery (2013). "The Missing 'One-offs': The Hidden Supply of High-Achieving, Low-Income Students." *Brookings Papers on Economic Activity*, Vol. 2013, pp. 1–65.

Hoxby, Caroline M., and Sarah Turner (2015). "What High-Achieving Low-Income Students Know About College." *The American Economic Review*, Vol. 105, pp. 514–17.

Hoxby, Caroline M., and Gretchen Weingarth (2005). "Taking Race Out of the Equation: School Reassignment and the Structure of Peer Effects." Unpublished.

Hu, Nan, Paul A. Pavlou, and Jennifer Zhang (2009). "Why Do Online Product Reviews Have a J-Shaped Distribution? Overcoming Biases in Online Word-of-Mouth Communication." *Communications of the ACM,* Vol. 52, pp. 144–47.

Huang, Jin (2013). "Intergenerational Transmission of Educational Attainment: The Role of Household Assets." *Economics of Education Review,* Vol. 33, pp. 112–23.

Hugonnier, Julien, Benjamin Lester, and Pierre-Olivier Weill (2014). "Heterogeneity in Decentralized Asset Markets." National Bureau of Economic Research Working Paper No. w20746.

Hung, Angela A., and Charles R. Plott (2001). "Information Cascades: Replication and an Extension to Majority Rule and Conformity-Rewarding Institutions." *The American Economic Review,* Vol. 91, pp. 1508–20.

Hwang, Victor W., and Greg Horowitt (2012). *The Rainforest: The Secret to Building the Next Silicon Valley.* San Francisco: Regenwald.

Ioannides, Y. M., and L. Datcher-Loury (2004). "Job Information Networks, Neighborhood Effects and Inequality." *Journal of Economic Literature,* Vol. 424, pp. 1056–93.

Jackson, Matthew O. (2003). "The Stability and Efficiency of Economic and Social Networks." In *Advances in Economic Design,* ed. S. Koray and M. Sertel. Heidelberg: Springer-Verlag.

—— (2007). "Social Structure, Segregation, and Economic Behavior." Nancy Schwartz Memorial Lecture, April 2007, Northwestern University. Available at SSRN: https://ssrn.com/abstract=1530885.

—— (2008). *Social and Economic Networks.* Princeton, N.J.: Princeton University Press.

—— (2008b). "Average Distance, Diameter, and Clustering in Social Networks with Homophily." In *The Proceedings of the Workshop in Internet and Network Economics (WINE 2008), Lecture Notes in Computer Science,* ed. C. Papadimitriou and S. Zhang. Berlin and Heidelberg: Springer-Verlag.

—— (2009). "Genetic Influences on Social Network Characteristics." *Proceedings of the National Academy of Sciences,* Vol. 106, pp. 1687–88.

—— (2014). "Networks in the Understanding of Economic Behaviors." *The Journal of Economic Perspectives,* Vol. 28 (4), pp. 3–22.

—— (2017). "A Typology of Social Capital and Associated Network Measures." Available at SSRN: http://ssrn.com/abstract=3073496.

—— (2018). "The Friendship Paradox and Systematic Biases in Perceptions and Social Norms," *Journal of Political Economy.*

Jackson, Matthew O., and Dunia López Pintado (2013). "Diffusion and Contagion in Networks with Heterogeneous Agents and Homophily." *Network Science,* Vol. 1 (1), pp. 49–67.

Jackson, Matthew O., and Stephen Nei (2015). "Networks of Military Alliances, Wars, and International Trade." *Proceedings of the National Academy of Sciences*, Vol. 112 (50), pp. 15277–84.

Jackson, Matthew O., Tomas R. Rodriguez Barraquer, and Xu Tan (2012). "Social Capital and Social Quilts: Network Patterns of Favor Exchange." *The American Economic Review*, Vol. 102, pp. 1857–97.

Jackson, Matthew O., and Brian W. Rogers (2007a). "Meeting Strangers and Friends of Friends: How Random Are Social Networks?" *The American Economic Review*, Vol. 97, pp. 890–915.

—— (2007b). "Relating Network Structure to Diffusion Properties Through Stochastic Dominance." *The BE Journal of Theoretical Economics*, Vol. 7, pp. 1–13.

Jackson, Matthew O., Brian W. Rogers, and Yves Zenou (2017). "The Economic Consequences of Social-Network Structure." *Journal of Economic Literature*, Vol. 55, pp. 49–95.

Jackson, Matthew O., and Evan C. Storms (2017). "Behavioral Communities and the Atomic Structure of Networks." ArXiv: 1710.04656.

Jackson, Matthew O., and Anne van den Nouweland (2005). "Strongly Stable Networks." *Games and Economic Behavior*, Vol. 51, pp. 420–44.

Jackson, Matthew O., and Asher Wolinsky (1996). "A Strategic Model of Social and Economic Networks." *Journal of Economic Theory*, Vol. 71, pp. 44–74.

Jackson, Matthew O., and Yiqing Xing (2014). "Culture-Dependent Strategies in Coordination Games." *Proceedings of the National Academy of Sciences*, Vol. 111 (3), pp. 10889–96.

—— (2018). "The Interaction of Communities, Religion, Governments, and Corruption in the Enforcement of Social Norms." Available at SSRN: https://ssrn.com/abstract=3153842.

Jackson, Matthew O., and Leeat Yariv (2011). "Diffusion, Strategic Interaction, and Social Structure." In *Handbook of Social Economics*, ed. Jess Benhabib, Alberto Bisin, and Matthew O. Jackson. San Diego: North-Holland.

Jackson, Matthew O., and Yves Zenou (2014). "Games on Networks." In *Handbook of Game Theory*, ed. H. P. Young and S. Zamir. Amsterdam: Elsevier.

Jadbabaie, Ali, Pooya Molavi, Alvaro Sandroni, Alireza Tahbaz-Salehi (2010). "Non-Bayesian Social Learning." *Games and Economic Behavior*, Vol. 76, pp. 210–25.

Jensen, Robert (2007). "The Digital Provide: Information (Technology), Market Performance, and Welfare in the South Indian Fisheries Sector." *The Quarterly Journal of Economics*, Vol. 122, pp. 879–924.

—— (2010). "The (Perceived) Returns to Education and the Demand for Schooling." *The Quarterly Journal of Economics*, Vol. 125 (2), pp. 515–48.

Jia, Ruixue, and Hongbin Li (2017). "Access to Elite Education, Wage Premium, and Social Mobility: The Truth and Illusion of China's College Entrance Exam." Working Paper, University of Toronto.

Johansen, Anders (2004). "Probing Human Response Times." *Physica A: Statistical Mechanics and Its Applications*, Vol. 338 (1), pp. 286–291.

Kaminsky, Graciela L., Carmen M. Reinhart, and Carlos A. Vegh (2003). "The Unholy Trinity of Financial Contagion." *The Journal of Economic Perspectives*, Vol. 17 (4), pp. 51–74.

Kaplan, Hillard, Kim Hill, Jane Lancaster, and A. Magdalena Hurtado (2000). "A Theory of Human Life History Evolution: Diet, Intelligence, and Longevity." *Evolutionary Anthropology: Issues, News, and Reviews*, Vol. 9, pp. 156–85.

Kaplan, Martin, and Matthew Hale (2010). "Local TV News in the Los Angeles Media Market: Are Stations Serving the Public Interest?" Norman Lear Center, USC Annenberg School for Communication & Journalism.

Kaplan, Steven N., and Joshua Rauh (2013). "It's the Market: The Broad-Based Rise in the Return to Top Talent." *The Journal of Economic Perspectives*, Vol. 27, pp. 35–55.

Karlan, Dean, and Valdivia, Martin (2011). "Teaching Entrepreneurship: Impact of Business Training on Microfinance Clients and Institutions." *Review of Economics and Statistics*, Vol. 93 (2), pp. 510–27.

Kasinitz, Philip, and Jan Rosenberg (1996). "Missing the Connection: Social Isolation and Employment on the Brooklyn Waterfront." *Social Problems*, Vol. 43, pp. 180–96.

Kasparov, Garry K., and Daniel King (2000). *Kasparov Against the World: The Story of the Greatest Online Challenge.* KasparovChess Online, Incorporated.

Katz, Elihu, and Paul F. Lazarsfeld (1955). *Personal Influence: The Part Played by People in the Flow of Mass Communication.* Glencoe, Ill.: Free Press.

Kaufmann, Katja Maria (2014). "Understanding the Income Gradient in College Attendance in Mexico: The Role of Heterogeneity in Expected Returns." *Quantitative Economics*, Vol. 5, pp. 583–630.

Kawachi, Ichiro, Daniel Kim, Adam Coutts, and S. V. Subramanian (2004). "Commentary: Reconciling the Three Accounts of Social Capital." *International Journal of Epidemiology*, Vol. 33, pp. 682–90.

Kearns, M. J., S. Judd, J. Tan, and J. Wortman (2009). "Behavioral Experiments on Biased Voting in Networks." *PNAS*, Vol. 106:5, pp. 1347–52.

Kelly, Morgan, and Cormac Ó Gráda (2000). "Market Contagion: Evidence from the Panics of 1854 and 1857." *The American Economic Review*, Vol. 90:5, pp. 1110–24.

Kempe, David, Jon Kleinberg, and Eva Tardos (2003). "Maximizing the Spread of Influence Through a Social Network." *Proceedings of the 9th International Conference on Knowledge Discovery and Data Mining*, pp. 137–46.

Kenney, Martin, ed. (2000). *Understanding Silicon Valley: The Anatomy of an Entrepreneurial Region.* Stanford, Calif.: Stanford University Press.

Kent, Dale (1978). *The Rise of the Medici: Faction in Florence, 1426–1434.* Oxford, U.K.: Oxford University Press.

Kets, Willemien, Garud Iyengar, Rajiv Sethi, and Samuel Bowles (2011). "Inequality and Network Structure." *Games and Economic Behavior,* Vol. 73, pp. 215–26.

Kets, Willemien, and Alvaro Sandroni (2016). "A Belief-Based Theory of Homophily." Available at SSRN: https://ssrn.com/abstract=2871514.

Keynes, John Maynard (1936). *The General Theory of Employment, Interest, and Money.* London: Macmillan.

Kim, David A., Alison R. Hwong, Derek Staff, D. Alex Hughes, A. James O'Malley, James H. Fowler, and Nicholas A. Christakis (2015). "Social Network Targeting to Maximise Population Behaviour Change: A Cluster Randomised Controlled Trial." *Lancet,* Vol. 386, pp. 145–53.

Kindelberger, Charles P., and Robert Z. Aliber (2000). *Manias, Panics and Crashes: A History of Financial Crises.* New York: Palgrave Macmillan.

Kinnan, Cynthia, and Robert Townsend (2012). "Kinship and Financial Networks, Formal Financial Access, and Risk Reduction." *The American Economic Review,* Vol. 102 (3), pp. 289–93.

Kivela, Mikko, Alexandre Arenas, James P. Gleeson, Yamir Moreno, and Mason A. Porter (2014). "Multilayer Networks." ArXiv:1309.7233v4 [physics.soc-ph].

Klasing, Mariko J., and Petros Milionis (2014). "Quantifying the Evolution of World Trade, 1870–1949." *Journal of International Economics,* Vol. 92, pp. 185–97.

Kleinberg, Jon M. (2000). "Navigation in a Small World." *Nature,* Vol. 406, p. 845.

Kleinberg, Jon M., Ravi Kumar, Prabhakar Raghavan, Sridhar Rajagopalan, and Andrew S. Tomkins (1999). "The Web as a Graph: Measurements, Models, and Methods." In *International Computing and Combinatorics Conference,* pp. 1–17. Heidelberg: Springer.

Klewes, Joachim, and Robert Wreschniok (2009). *Reputation Capital.* Heidelberg: Springer.

Kloumann, Isabel, Lada Adamic, Jon Kleinberg, and Shaomei Wu (2015). "The Lifecycles of Apps in a Social Ecosystem." In *Proceedings of the 24th International Conference on World Wide Web,* pp. 581–91.

Knack, Stephen, and Philip Keefer (1997). "Does Social Capital Have an Economic Payoff? A Cross-Country Investigation." *The Quarterly Journal of Economics,* Vol. 112, pp. 1251–88.

Knight, Frank H. (1924). "Some Fallacies in the Interpretation of Social Cost." *The Quarterly Journal of Economics,* Vol. 38 (4), pp. 582–606.

Kohn, George C., ed. (2008). *Encyclopedia of Plague and Pestilence: From Ancient Times to the Present.* Infobase Publishing.

König, Michael D., Claudio J. Tessone, and Yves Zenou (2014). "Nestedness in Networks: A Theoretical Model and Some Applications." *Theoretical Economics*, Vol. 9, pp. 695–752.

Krackhardt, David (1987). "Cognitive Social Structures." *Social Networks*, Vol. 9, pp. 109–34.

—— (1996). "Structural Leverage in Marketing." In Dawn Iacobucci, ed., *Networks in Marketing*. pp. 50–59. Thousand Oaks, CA: Sage.

Krapivsky, Paul L., Sidney Redner, and Francois Leyvraz (2000). "Connectivity of Growing Random Networks." *Physical Review Letters*, Vol. 85, p. 4629.

Kremer, Michael, Nazmul Chaudhury, F. Halsey Rogers, Karthik Muralidharan, and Jeffrey Hammer (2005). "Teacher Absence in India: A Snapshot." *Journal of the European Economic Association*, Vol. 3, pp. 658–67.

Krugman, Paul (2014). "Why We're in a New Gilded Age." *The New York Review of Books*, May 8.

Kuhn, Peter, Peter Kooreman, Adriaan Soetevent, and Arie Kapteyn (2011). "The Effects of Lottery Prizes on Winners and Their Neighbors: Evidence from the Dutch Postcode Lottery." *American Economic Review*, Vol. 101 (5), pp. 2226–47.

Kumar, Ravi, Prabhakar Raghavan, Sridhar Rajagopalan, D. Sivakumar, Andrew Tomkins, and Eli Upfal (2000). "Stochastic Models for the Web Graph." In *Foundations of Computer Science, 2000*, Proceedings, 41st Annual Symposium, pp. 57–65, IEEE.

Laguna Müggenburg, Eduardo (2017). "Using Homophily to Spread Information: Influencers Need Not Be Superstars." Mimeo, Stanford University.

Lalanne, Marie, and Paul Seabright (2016). "The Old Boy Network: The Impact of Professional Networks on Remuneration in Top Executive Jobs." SAFE Working Paper No 123.

Lareau, Annette (2011). *Unequal Childhoods: Class, Race, and Family Life*. Berkeley: University of California Press.

Laschever, Ron A. (2011). "The Doughboys Network: Social Interactions and Labor Market Outcomes of World War I Veterans." SSRN Discussion Paper No. 1205543.

Lau, Max S. Y., Benjamin Douglas Dalziel, Sebastian Funk, Amanda McClelland, Amanda Tiffany, Steven Riley, C. Jessica E. Metcalf, and Bryan T. Grenfell (2017). "Spatial and Temporal Dynamics of Superspreading Events in the 2014–2015 West Africa Ebola Epidemic." *Proceedings of the National Academy of Sciences*, Vol. 114 (9), pp. 2337–42.

Lazarsfeld, Paul F., Bernard Berelson, and Hazel Gaudet (1948). *The People's Choice: How the Voter Makes Up His Mind in a Presidential Campaign*. New York: Columbia University Press.

Lazarsfeld, Paul F., and Robert K. Merton (1954). "Friendship as a Social Process: A Substantive and Methodological Analysis." In *Freedom and Control in Modern Society*, ed. M. Berger. New York: Van Nostrand.

Lazear, Edward P. (1999). "Culture and Language." *Journal of Political Economy*, Vol. 107, pp. S95–S126.

Lazer, David, and Stefan Wojcik (2017). "Political Networks and Computational Social Science." In *The Oxford Handbook of Political Networks*, ed. Jennifer Victor, Alexander Montgomery, and Mark Lubell. New York: Oxford University Press.

Leduc, Matt V., Matthew O. Jackson, and Ramesh Johari (2016). "Pricing and Referrals in Diffusion on Networks." SSRN Paper No. 2425490.

Lee, Chu Lin, and Gary Solon (2009). "Trends in Intergenerational Income Mobility." *Review of Economics and Statistics*. Vol. 91 (4), pp. 766–72.

Lelkes, Yphtach, Gaurav Sood, and Shanto Iyengar (2015). "The Hostile Audience: The Effect of Access to Broadband Internet on Partisan Affect." *American Journal of Political Science*, Vol. 61 (1), pp. 5–20.

Lerman, Kristina, Xiaoran Yan, and Xin-Zeng Wu (2015). "The Majority Illusion in Social Networks." ArXiv: 1506.03022v1.

Lerner, Josh, and Ulrike Malmendier (2013). "With a Little Help from My (Random) Friends: Success and Failure in Post–Business School Entrepreneurship." *Review of Financial Studies*, Vol. 26, pp. 2411–52.

Levitt, Steven D., and Stephen J. Dubner (2005). *Freakonomics*. New York: HarperCollins.

Levy, Gilat, and Ronny Razin (2015). "Correlation Neglect, Voting Behavior, and Information Aggregation." *The American Economic Review*, Vol. 105, pp. 1634–45.

Li, Weihua, Aisha E. Bradshaw, Caitlin B. Clary, and Skyler J. Cranmer (2017). "A Three-Degree Horizon of Peace in the Military Alliance Network," *Science Advances*, Vol. 3, article no. e1601895.

Lin, Ken-Hou, and Jennifer Lundquist (2013). "Mate Selection in Cyberspace: The Intersection of Race, Gender, and Education." *The American Journal of Sociology*, Vol. 119 (1), pp. 183–215.

Lin, Nan (1999). "Building a Network Theory of Social Capital." *Connections*, Vol. 22, pp. 28–51.

Lindert, Peter H., and Jeffrey G. Williamson (2012). "American Incomes 1774–1860." National Bureau of Economic Research Working Paper No. 18396.

Lindquist, Matthew J., and Yves Zenou (2014). "Key Players in Co-offending Networks." Available at SSRN: http://ssrn.com/abstract=2444910.

List, John A. (2006). "The Behavioralist Meets the Market: Measuring Social Preferences and Reputation Effects in Actual Transactions." *Journal of Political Economy*, Vol. 114, pp. 1–37.

List, John A., and Imran Rasul (2011). "Field Experiments in Labor Economics." *Handbook of Labor Economics*, Vol. 4, pp. 103–228.

Lobel, Ilan, and Evan Sadler (2015). "Information Diffusion in Networks Through Social Learning." *Theoretical Economics*, Vol. 10, pp. 807–51.

Lorrain, François, and Harrison C. White (1971). "Structural Equivalence of Individuals in Social Networks." *The Journal of Mathematical Sociology,* Vol. 1, pp. 49–80.

Loury, Glenn (1977). "A Dynamic Theory of Racial Income Differences." *Women, Minorities, and Employment Discrimination,* Vol. 153, pp. 86–153.

—— (2009). *The Anatomy of Racial Inequality.* Cambridge, Mass.: Harvard University Press.

Lucas, Robert E., Jr. (2013). "Glass-Steagall: A Requiem." *The American Economic Review: Papers and Proceedings,* Vol. 103 (3), pp. 43–47.

Ludwig, Jens, Jeffrey B. Liebman, Jeffrey R. Kling, Greg J. Duncan, Lawrence F. Katz, Ronald C. Kessler, and Lisa Sanbonmatsu (2008). "What Can We Learn About Neighborhood Effects from the Moving to Opportunity Experiment." *American Journal of Sociology,* Vol. 114, pp. 144–88.

Mailath, George J., and Larry Samuelson (2006). *Repeated Games and Reputations: Long-Run Relationships.* Oxford, U.K.: Oxford University Press.

Manski, Charles F. (1993). "Identification of Endogenous Social Effects: The Reflection Problem." *The Review of Economic Studies,* pp. 531–42.

—— (2006). "Interpreting the Predictions of Prediction Markets." *Economics Letters,* Vol. 91, pp. 425–29.

Marmaros, David, and Bruce Sacerdote (2002). "Peer and Social Networks in Job Search." *European Economic Review,* Vol. 46, pp. 870–79.

Marshall, Alfred (1890). *Principles of Political Economy.* New York: Macmillan.

Martinelli, César A., Susan Parker, Ana Cristina Pérez-Gea, and Rodimiro Rodrigo (2015). "Cheating and Incentives: Learning from a Policy Experiment." Available at SSRN: https://ssrn.com/abstract=2606487.

—— (2016). "Cheating and Incentives: Learning from a Policy Experiment." Working Paper, George Mason University.

Marvel, Seth A., Travis Martin, Charles R. Doering, David Lusseau, and Mark E. J. Newman (2013). "The Small-World Effect Is a Modern Phenomenon." ArXiv: 1310.2636.

Massey, Douglas S., Joaquin Arango, Graeme Hugo, Ali Kouaouci, Adela Pellegrino, and J. Edward Taylor (1998). *Worlds in Motion: Understanding International Migration at the End of the Millennium.* Oxford, U.K.: Oxford University Press.

Massey, Douglas S., and Nancy A. Denton (1993). *American Apartheid: Segregation and the Making of the Underclass.* Cambridge, Mass.: Harvard University Press.

Mauleon, Ana, and Vincent Vannetelbosch (2004). "Farsightedness and Cautiousness in Coalition Formation Games with Positive Spillovers." *Theory and Decision,* Vol. 56, pp. 291–324.

Mazzocco, Maurizio, and Shiv Saini (2012). "Testing Efficient Risk Sharing with Heterogeneous Risk Preferences." *The American Economic Review,* Vol. 102, pp. 428–68.

McCarty, Christopher, Peter D. Killworth, H. Russell Bernard, and Eugene C. Johnsen (2001). "Comparing Two Methods for Estimating Network Size." *Human Organization*, Vol. 60 (1), pp. 28–39.

McCormick, Tyler H., Matthew J. Salganik, and Tian Zheng (2010). "How Many People Do You Know?: Efficiently Estimating Personal Network Size." *Journal of the American Statistical Association*, Vol. 105 (489), pp. 59–70.

McCoy, Elin (2014). *The Emperor of Wine: The Rise of Robert M. Parker, Jr., and the Reign of American Taste*. New York: HarperCollins.

McFarland, Daniel A., James Moody, David Diehl, Jeffrey A. Smith, and Reuben J. Thomas (2014). "Network Ecology and Adolescent Social Structure." *American Sociological Review*, Vol. 79 (6), pp. 1088–1121.

McPherson, Miller, Lynn Smith-Lovin, and James M. Cook (2001). "Birds of a Feather: Homophily in Social Networks." *Annual Review of Sociology*, Vol. 27 (1), pp. 415–44.

McShane, Blakeley B., Eric T. Bradlow, and Jonah Berger (2012). "Visual Influence and Social Groups." *Journal of Marketing Research*, Vol. 49 (6), pp. 854–71.

Mele, Angelo (2017). "A Structural Model of Segregation in Social Networks." *Econometrica*, Vol. 85 (3), pp. 825–50.

Mengel, Friederike (2015). "Gender Differences in Networking." Available at SSRN: https://ssrn.com/abstract=2636885.

Miller, John H., and Scott E. Page (2009). *Complex Adaptive Systems: An Introduction to Computational Models of Social Life*. Princeton, N.J.: Princeton University Press.

Mitzenmacher, Michael (2004). "A Brief History of Generative Models for Power Law and Lognormal Distributions," *Internet Mathematics*, Vol. 1, pp. 226–51.

Mobius, Markus, Tuan Phan, and Adam Szeidl (2015). "Treasure Hunt: Social Learning in the Field." National Bureau of Economic Research Paper No. 21014.

Molavi, Pooya, Alireza Tahbaz-Salehi, and Ali Jadbabaie (2018). "A Theory of Non-Bayesian Social Learning." *Econometrica*, Vol. 86 (2), pp. 445–90.

Monsted, Bjarke, Piotr Sapiezynski, Emilio Ferrara, and Sune Lehmann (2017). "Evidence of Complex Contagion of Information in Social Media: An Experiment Using Twitter Bots." ArXiv preprint arXiv:1703.06027.

Montgomery, James D. (1991). "Social Networks and Labor-Market Outcomes: Toward an Economic Analysis." *The American Economic Review*, Vol. 81, pp. 1408–18.

Moody, James, and Peter J. Mucha (2013). "Portrait of Political Party Polarization." *Network Science*, Vol. 1 (1), pp. 119–21.

Moore, Cristopher, and Mark E. J. Newman (2000). "Epidemics and Percolation in Small-World Networks." *Physical Review E*, Vol. 61, p. 5678.

Morelli, Sylvia A., Desmond C. Ong, Rucha Makati, Matthew O. Jackson, and Jamil Zaki (2017). "Psychological Trait Correlates of Individuals'

Positions in Social Networks." *Proceedings of the National Academy of Sciences*, Vol. 114 (37), pp. 9843–47.

Moretti, Enrico (2011). "Social Learning and Peer Effects in Consumption: Evidence from Movie Sales." *The Review of Economic Studies*, Vol. 78, pp. 356–93.

Morris, Stephen (2000). "Contagion." *Review of Economic Studies*, Vol. 67 (1), pp. 57–78.

Munshi, Kaivan (2003). "Networks in the Modern Economy: Mexican Migrants in the US Labor Market." *The Quarterly Journal of Economics*, Vol. 118, pp. 549–99.

Muralidharan, Karthink, and Michael Kremer (2009). *Public and Private Schools in Rural India*, pp. 1–27. Cambridge, Mass.: MIT Press.

Myers, Charles A., and George P. Shultz (1951). *The Dynamics of a Labor Market*. Chicago: Greenwood.

Nannicini, Tommaso, Andrea Stella, Guido Tabellini, and Ugo Troiano (2013). "Social Capital and Political Accountability." *The American Economic Journal: Economic Policy*, Vol. 5, pp. 222–50.

Neal, Larry D., and Marc D. Weidenmier (2003). "Crises in the Global Economy from Tulips to Today." In *Globalization in Historical Perspective*, pp. 473–514. Chicago: University of Chicago Press.

Nei, Stephen M. (2017). "Frictions in Information Aggregation in Social Learning Environments." Mimeo. Oxford University.

Neuman, Susan B., and Donna Celano (2012). *Giving Our Children a Fighting Chance: Poverty, Literacy, and the Development of Information Capital*. New York: Teachers College Press.

Newman, Mark E. J. (2003). "The Structure and Function of Complex Networks." *SIAM Review*, Vol. 45 (2), pp. 167–256.

Newman, Mark E. J., Cristopher Moore, and Duncan J. Watts (2000). "Mean-Field Solution of the Small-World Network Model." *Physical Review Letters*, Vol. 84, pp. 3201–04.

Nguyen, Trang (2008). "Information, Role Models and Perceived Returns to Education: Experimental Evidence from Madagascar." Manuscript, MIT.

Ody-Brasier, Amandine, and Isabel Fernandez-Mateo (2015). "Minority Producers and Pricing in the Champagne Industry: The Case of Female Grape Growers." *Academy of Management Proceedings*, Vol. 2015 (1), pp. 124–57.

Ostrom, Elinor (1990). *Governing the Commons: The Evolution of Institutions for Collective Action*. Cambridge, U.K.: Cambridge University Press.

Ozsoylev, Han N., Johan Walden, M. Deniz Yavuz, and Recep Bildik (2014). "Investor Networks in the Stock Market." *The Review of Financial Studies*, Vol. 27 (5). pp. 1323–66.

Padgett, John F., and Christopher K. Ansell (1993). "Robust Action and the Rise of the Medici 1400–1434." *American Journal of Sociology*, Vol. 98 (6), pp. 1259–1319.

Page, Scott E. (2008). *The Difference: How the Power of Diversity Creates Better Groups, Firms, Schools, and Societies.* Princeton, N.J.: Princeton University Press.

——— (2017). *The Diversity Bonus: How Great Teams Pay Off in the Knowledge Economy.* Princeton, N.J.: Princeton University Press.

Pallais, Amanda, and Emily Glassberg Sands (2016). "Why the Referential Treatment: Evidence from Field Experiments on Referrals." *Journal of Political Economy,* Vol. 124 (6), pp. 1793–1828.

Patacchini, Eleonora, and Yves Zenou (2012). "Ethnic Networks and Employment Outcomes." *Regional Science and Urban Economics,* Vol. 42, pp. 938–49.

Payne, B. Keith, Jazmin L. Brown-Iannuzzi, and Jason W. Hannay (2017). "Economic Inequality Increases Risk Taking." *Proceedings of the National Academy of Sciences,* Vol. 114, pp. 4643–48.

Pennock, David M., Gary W. Flake, Steve Lawrence, Eric J. Glover, and C. Lee Giles (2002). "Winners Don't Take All: Characterizing the Competition for Links on the Web." *Proceedings of the National Academy of Sciences,* Vol. 99, pp. 5207–11.

Perkins, H. Wesley, Michael P. Haines, and Richard Rice (2005). "Misperceiving the College Drinking Norm and Related Problems: A Nationwide Study of Exposure to Prevention Information, Perceived Norms and Student Alcohol Misuse." *Journal of Studies on Alcohol,* Vol. 66, pp. 470–78.

Perkins, H. Wesley, Jeffrey W. Linkenbach, Melissa A. Lewis, and Clayton Neighbors (2010). "Effectiveness of Social Norms Media Marketing in Reducing Drinking and Driving: A Statewide Campaign." *Addictive Behaviors,* Vol. 35, pp. 866–74.

Perkins, H. Wesley, Philip W. Meilman, Jami S. Leichliter, Jeffrey R. Cashin, and Cheryl A. Presley (1999). "Misperceptions of the Norms for the Frequency of Alcohol and Other Drug Use on College Campuses." *Journal of American College Health,* Vol. 47:6, pp. 253–58.

Persson, Torsten, and Guido Tabellini (1994). "Is Inequality Harmful for Growth?" *The American Economic Review,* Vol. 84, pp. 600–621.

Pfeffer, Jeffrey, (1981). *Power in Organizations,* Vol. 33, Marshfield, Mass.: Pitman.

Pfeffer, Jeffrey (1992). *Managing with Power: Politics and Influence in Organizations.* Cambridge, Mass.: Harvard Business Press.

Pfitzner, René, Ingo Scholtes, Antonios Garas, Claudio J. Tessone, and Frank Schweitzer (2013). "Betweenness Preference: Quantifying Correlations in the Topological Dynamics of Temporal Networks." *Physical Review Letters,* Vol. 110, article no. 198701.

Piketty, Thomas (2014). *Capital in the Twenty-First Century.* Cambridge, Mass.: Harvard University Press.

Popkin, Susan J., Laura E. Harris, and Mary K. Cunningham (2002). "Families in Transition: A Qualitative Analysis of the MTO Experience."

Report Prepared for the U.S. Department of Housing and Urban Development.

Porter, Mason A., Jukka-Pekka Onnela, and Peter J. Mucha (2009). "Communities in Networks." *Notices of the AMS*, Vol. 56, pp. 1082–97.

Poy, Samuele, and Simone Schüller (2016). "Internet and Voting in the Web 2.0 Era: Evidence from a Local Broadband Policy." CESIFO Working Paper No. 6129.

Prabhakar, Balaji, Katherine N. Dektar, and Deborah M. Gordon (2012). "The Regulation of Ant Colony Foraging Activity Without Spatial Information." *PLoS Computational Biology*, Vol. 8, article no. e1002670.

Prendergast, Canice, and Robert H. Topel (1996). "Favoritism in Organizations." *Journal of Political Economy*, Vol. 104, pp. 958–78.

Price, Derek de Solla (1976). "A General Theory of Bibliometric and Other Cumulative Advantage Processes." *Journal of the American Society for Information Science*, Vol. 27, pp. 292–306.

Prummer, Anja (2016). "Spatial Advertisement in Political Campaigns." Preprint, Queen Mary University of London.

Putnam, R. D. (2000). *Bowling Alone: The Collapse and Revival of American Community*. New York: Simon & Schuster.

Quillian, Lincoln (1995). "Prejudice as a Response to Perceived Group Threat: Population Composition and Anti-Immigrant and Racial Prejudice in Europe," *American Sociological Review*, pp. 586–611.

Rainie, Lee, and Barry Wellman (2012). *Networked: The New Social Operating System*. Cambridge, Mass.: MIT Press.

Ravallion, Martin, and Shubham Chaudhuri (1997). "Risk and Insurance in Village India: Comment." *Econometrica*, Vol. 65:1, pp. 171–84.

Rawls, John (1971). *A Theory of Justice*. Cambridge, Mass.: Harvard University Press.

Reardon, Sean F. (2011). "The Widening Academic Achievement Gap Between the Rich and the Poor: New Evidence and Possible Explanations." In *Whither Opportunity*, pp. 91–116. New York: Russell Sage Foundation.

Rees, Albert, George Pratt Shultz, et al. (1970). *Workers and Wages in an Urban Labor Market*. Chicago: University of Chicago Press.

Reinhart, Carmen, and Kenneth Rogoff (2009). *This Time Is Different*. Princeton, N.J.: Princeton University Press.

Revzina, N. V., and R. J. DiClemente (2005). "Prevalence and Incidence of Human Papillomavirus Infection in Women in the USA: A Systematic Review." *International Journal of STD and AIDS*, Vol. 16, pp. 528–37.

Riach, Peter A., and Judith Rich (2002). "Field Experiments of Discrimination in the Marketplace." *The Economic Journal*, Vol. 112, pp. F480–F518.

Richerson, Peter J., and Robert Boyd (2008). *Not by Genes Alone: How Culture Transformed Human Evolution*. Chicago: University of Chicago Press.

Roemer, John E., and Alain Trannoy (2016). "Equality of Opportunity: Theory and Measurement." *Journal of Economic Literature*, Vol. 54, pp. 1288–1332.

Rogers, Everett M. (1995). *Diffusion of Innovations*. New York: Free Press.

Ross, Lee, David Greene, and Pamela House (1977). "The False Consensus Effect: An Ego-Centric Bias in Social Perception and Attribution Processes." *Journal of Experimental Social Psychology*, Vol. 13, pp. 279–301.

Roth, Alvin E. (2007). "Repugnance as a Constraint on Markets." *The Journal of Economic Perspectives*, Vol. 21, pp. 37–58.

Sacerdote, Bruce (2001). "Peer Effects with Random Assignment: Results for Dartmouth Roommates." *The Quarterly Journal of Economics*, Vol. 116, pp. 681–704.

——— (2011). "Peer Effects in Education: How Might They Work, How Big Are They and How Much Do We Know Thus Far?" In *Handbook of the Economics of Education*, ed. by Eric A. Hanushek, Stephen Machin, and Ludger Woessmann, Vol. 3, pp. 249–77. San Diego: North-Holland.

Salganik, Matthew J., Peter S. Dodds, and Duncan J. Watts (2006). "Experimental Study of Inequality and Unpredictability in an Artificial Cultural Market." *Science*, Vol. 311, pp. 854–56.

Samphantharak, Krislert, and Robert M. Townsend (2018). "Risk and Return in Village Economies." *American Economic Journal: Microeconomics*, Vol. 10 (1), pp. 1–40.

Saunders, Anthony, and Berry Wilson (1996). "Contagious Bank Runs: Evidence from the 1929–1933 Period." *Journal of Financial Intermediation*, Vol. 5, pp. 409–23.

Schaner, Simone (2015). "Do Opposites Detract? Intrahousehold Preference Heterogeneity and Inefficient Strategic Savings." *American Economic Journal: Applied Economics*, Vol. 7 (2), pp. 135–74.

Scharfstein, David S., and Jeremy C. Stein (1990). "Herd Behavior and Investment." *The American Economic Review*, pp. 465–79.

Schmitt, Robert C., and Eleanor C. Nordyke (2001). "Death in Hawai'i: The Epidemics of 1848–1849." *The Hawaiian Journal of History*, Vol. 35, pp. 1–13.

Schweitzer, Frank, Giorgio Fagiolo, Didier Sornette, Fernando Vega-Redondo, Alessandro Vespignani, and Douglas R. White (2009). "Economic Networks: The New Challenges." *Science*, Vol. 325, pp. 422–25.

Scitovsky, Tibor (1954). "Two Concepts of External Economies." *Journal of Political Economy*, Vol. 62 (2), pp. 143–51.

Seabright, Paul (2010). *The Company of Strangers: A Natural History of Economic Life*. Princeton, N.J.: Princeton University Press.

Shemesh, Joshua, and Fernando Zapatero (2016). "The Intensity of Keeping Up with the Joneses Behavior: Evidence from Neighbor Effects in Car Purchases." Preprint.

Shi, Yigong, and Yi Rao (2010). "China's Research Culture." *Science*, Vol. 329 (5996), p. 1128.

Shiller, Robert J. (2015). *Irrational Exuberance*. Princeton, N.J.: Princeton University Press.

Shilts, Randy (1987). *And the Band Played On: Politics, People, and the AIDS Epidemic*. New York: St. Martin's Press.

Shoag, Daniel, and Stan Veuger (2016). "No Woman No Crime: Ban the Box, Employment, and Upskilling." HKS Working Paper No. 16-015.

Shulman, Stanford T., Deborah L. Shulman, and Ronald H. Sims (2009). "The Tragic 1824 Journey of the Hawaiian King and Queen to London: History of Measles in Hawaii." *Pediatric Infectious Disease Journal*, Vol. 28 (8), pp. 728–33.

Skopek, Jan, Florian Schulz, and Hans-Peter Blossfeld (2010). "Who Contacts Whom? Educational Homophily in Online Mate Selection." *European Sociological Review*, Vol. 27 (2), pp. 180–195.

Smith, Adam (1776). *An Inquiry into the Nature and Causes of the Wealth of Nations*. London: A. and C. Black and W. Tait.

Smith, Lones, and Peter Sorensen (2000). "Pathological Outcomes of Observational Learning." *Econometrica*, Vol. 68, pp. 371–98.

Smith, Sandra S. (2000). "Mobilizing Social Resources: Race, Ethnic, and Gender Differences in Social Capital and Persisting Wage Inequalities." *The Sociological Quarterly*, Vol. 41 (4), pp. 509–37.

Snijders, Tom A. B. (2011). "Multilevel Analysis." In *International Encyclopedia of Statistical Science*, pp. 879–82. New York: Springer.

Sobel, Joel (2002). "Can We Trust Social Capital?" *Journal of Economic Literature*, Vol. 40, pp. 139–54.

Solomonoff, Ray, and Anatol Rapoport (1951). "Connectivity of Random Nets." *The Bulletin of Mathematical Biophysics*, Vol. 13, pp. 107–17.

Solow, Robert M. (2000). "Notes on Social Capital and Economic Performance." In *Social Capital: A Multifaceted Perspective*, Vol. 6, pp. 6–12. Washington, D.C.: World Bank Publications.

Soramäki, Kimmo, Morten L. Bech, Jeffrey Arnold, Robert J. Glass, and Walter E. Beyeler (2007). "The Topology of Interbank Payment Flows." *Physica A*, Vol. 379, pp. 317–33.

Stanley, H. Eugene (1971). *Phase Transitions and Critical Phenomena*. Oxford, U.K.: Oxford University Press.

Stock, James H., and Francesco Trebbi (2003). "Retrospectives: Who Invented Instrumental Variable Regression?" *The Journal of Economic Perspectives*, Vol. 17, pp. 177–94.

Su, Lixin Nancy, and Donghui Wu (2016). "Is Audit Behavior Contagious? Teamwork Experience and Audit Quality by Individual Auditors." Available at SSRN: https://ssrn.com/abstract=2816435.

Sunstein, Cass R. (2108). *Republic: Divided Democracy in the Age of Social Media*. Princeton, N. J.: Princeton University Press.

Surowiecki, James (2005). *The Wisdom of Crowds*. New York: Anchor.

Szreter, Simon, and Michael Woolcock (2004). "Health by Association?

Social Capital, Social Theory, and the Political Economy of Public Health." *International Journal of Epidemiology,* Vol. 33, pp. 650–67.

Tabellini, Guido (2010). "Culture and Institutions: Economic Development in the Regions of Europe." *Journal of the European Economic Association,* Vol. 8, pp. 677–716.

Tadelis, Steven (2016). "The Economics of Reputation and Feedback Systems in E-Commerce Marketplaces." *IEEE Internet Computing,* Vol. 20 (1), 12–19.

Tan, Jijun, Ting Zeng, and Shenghao Zhu (2015). "Earnings, Income, and Wealth Distributions in China: Facts from the 2011 China Household Finance Survey." Preprint.

Tan, Tongxue (2016). *Social Ties and the Market: A Study of Digital Printing Industry from an Informal Economy Perspective,* Chapter 3. Beijing: BRILL.

Tatum, Beverly Daniel (2010). *Why Are All the Black Kids Sitting Together in the Cafeteria? And Other Conversations About Race.* New York: Basic Books.

Thaler, Richard H., and Cass R. Sunstein (2008). *Nudge: Improving Decisions About Health, Wealth, and Happiness.* New Haven, Conn.: Yale University Press.

Tomasello, Michael (2009). *The Cultural Origins of Human Cognition.* Cambridge, Mass.: Harvard University Press.

Townsend, Robert M. (1994). "Risk and Insurance in Village India." *Econometrica,* Vol. 62, pp. 539–91.

Travers, Jeffrey, and Stanley Milgram (1969). "An Experimental Study of the Small World Problem." *Sociometry,* Vol. 32 (4), pp. 425-43.

Tucker, Joan S., Jeremy N. V. Miles, Elizabeth J. D'Amico, Annie J. Zhou, Harold D. Green, and Regina A. Shih (2013). "Temporal Associations of Popularity and Alcohol Use Among Middle School Students." *Journal of Adolescent Health,* Vol. 52, pp. 108–15.

Ugander, Johan, Brian Karrer, Lars Backstrom, and Cameron Marlow (2011). "The Anatomy of the Facebook Social Graph." http://arxiv.org/abs/1111.4503v1.

Uzzi, Brian (1996). "The Sources and Consequences of Embeddedness for the Economic Performance of Organizations: The Network Effect." *American Sociological Review,* pp. 674–98.

—— (1997). "Social Structure and Competition in Interfirm Networks: The Paradox of Embeddedness." *Administrative Science Quarterly,* Vol. 42, pp. 35–67.

Valente, Thomas W. (2012). "Network Interventions." *Science,* Vol. 337 (6), pp. 49–53.

Valente, Thomas W., and Patchareeya Pumpuang (2007). "Identifying Opinion Leaders to Promote Behavior Change." *Health Education and Behavior,* Vol. 34 (6), pp. 881–96.

Valente, Thomas W., Jennifer B. Unger, and C. Anderson Johnson (2005). "Do Popular Students Smoke? The Association Between Popularity and

Smoking Among Middle School Students." *Journal of Adolescent Health,* Vol. 37, pp. 323–29.

Valentine, Vikki (2006). "Origins of the 1918 Pandemic: The Case for France." NPR, http://www.npr.org/templates/story/story.php?storyId=5222069.

van der Leij, Marco, Daan in 't Veld, and Cars H. Hommes (2016). "The Formation of a Core-Periphery Structure in Heterogeneous Financial Networks." Available at SSR: https://ssrn.com/abstract=2865666.

Vázquez, Alexei (2003). "Growing Network with Local Rules: Preferential Attachment, Clustering Hierarchy, and Degree Correlations." *Physical Review E,* Vol. 67, article no. 056104.

Vega-Redondo, Fernando (2007). *Complex Social Networks.* Cambridge, U.K.: Cambridge University Press.

Verbrugge, Lois M. (1977). "The Structure of Adult Friendship Choices." *Social Forces,* Vol. 56, pp. 576–97.

Wainwright, Tom (2016). *Narconomics: How to Run a Drug Cartel.* New York: PublicAffairs.

Waldman, Steven, and the Working Group on Information Needs of Communities (2011). *Information Needs of Communities: The Changing Media Landscape in a Broadband Age.* Federal Communications Commission Report, http://www.fcc.gov/infoneedsreport.

Wang, Chaojun (2017). "Core-Periphery Trading Networks." Dissertation, Stanford University.

Wartick, Steven L. (2002). "Measuring Corporate Reputation: Definition and Data." *Business & Society,* Vol. 41, pp. 371–92.

Wasserman, Stanley, and Katherine Faust (1994). *Social Network Analysis.* Cambridge, U.K.: Cambridge University Press.

Watts, Alison (2001). "A Dynamic Model of Network Formation." *Games and Economic Behavior,* Vol. 34, pp. 331–41.

Watts, Duncan J. (1999). *Small Worlds: The Dynamics of Networks Between Order and Randomness.* Princeton, N.J.: Princeton University Press.

——— (2004). *Six Degrees: The Science of a Connected Age.* New York: W. W. Norton.

Watts, Duncan J., and Steven H. Strogatz (1998). "Collective Dynamics of Small-World Networks." *Nature,* Vol. 393, pp. 440–42.

Weichselbaumer, Doris, and Rudolf Winter-Ebmer (2005). "A Meta-Analysis of the International Gender Wage Gap." *Journal of Economic Surveys,* Vol. 19, pp. 479–511.

Weisel, Ori, and Shaul Shalvi (2015). "The Collaborative Roots of Corruption." *Proceedings of the National Academy of Sciences,* Vol. 112, pp. 10651–56.

Wellman, Barry, and Stephen D. Berkowitz (1988). *Social Structures: A Network Approach.* Cambridge, U.K.: Cambridge University Press.

West, Cornel (1993). *Race Matters.* New York: Vintage.

White, Douglas R., and Karl P. Reitz (1983). "Graph and Semigroup Homomorphisms on Networks of Relations." *Social Networks,* Vol. 5, pp. 193–234.

Wilfert, L., G. Long, H. C. Leggett, P. Schmid-Hempel, R. Butlin, S. J. M. Martin, and M. Boots (2016). "Deformed Wing Virus Is a Recent Global Epidemic in Honeybees Driven by Varroa Mites." *Science,* Vol. 351, pp. 594–97.

Willette, Demian A., Sara E. Simmonds, Samantha H. Cheng, Sofi Esteves, Tonya L. Kane, Hayley Nuetzel, Nicholas Pilaud, Rita Rachmawati, and Paul H. Barber (2017). "Using DNA Barcoding to Track Seafood Mislabeling in Los Angeles Restaurants." *Conservation Biology.* Vol. 31 (5), pp. 1076–85.

Wilson, William Julius (2012). *The Truly Disadvantaged: The Inner City, the Underclass, and Public Policy.* Chicago: University of Chicago Press.

Wolfers, Justin, and Eric Zitzewitz (2004). "Prediction Markets." *The Journal of Economic Perspectives,* Vol. 18, pp. 107–26.

—— (2006). "Interpreting Prediction Market Prices as Probabilities." National Bureau of Economic Research Paper No. 12200.

Woolcock, Michael (1998). "Social Capital and Economic Development: Toward a Theoretical Synthesis and Policy Framework." *Theory and Society,* Vol. 27, pp. 151–208.

Wu, Ye, Changsong Zhou, Jinghua Xiao, Jürgen Kurths, and Hans Joachim Schellnhuber (2010). "Evidence for a Bimodal Distribution in Human Communication." *Proceedings of the National Academy of Sciences,* Vol. 107 (44), pp. 18803–8.

Xie, Yu, and Xiang Zhou (2014). "Income Inequality in Today's China." *Proceedings of the National Academy of Sciences,* Vol. 111, pp. 6928–33.

Xing, Yiqing (2016). "Who Shares Risk with Whom and How? Endogenous Matching and Selection of Risk Sharing Equilibria." SIEPR Discussion Paper No. 16-025.

Young, Allyn A. (1913). "Review of Pigou's Wealth and Welfare." *The Quarterly Journal of Economics,* Vol. 27 (4), pp. 672–86

Young, H. Peyton (1996). "The Economics of Convention." *The Journal of Economic Perspectives,* Vol. 10, pp. 105–22.

Zuckoff, Mitchell (2006). *Ponzi's Scheme: The True Story of a Financial Legend.* New York: Random House.

# INDEX

NOTE: Page numbers followed by *f* indicate a figure.

A NOTE ON THE TYPE

This book was set in Minion, a typeface produced by the Adobe Corporation specifically for the Macintosh personal computer, and released in 1990. Designed by Robert Slimbach, Minion combines the classic characteristics of old style faces with the full complement of weights required for modern typesetting.

*Composed by North Market Street Graphics, Lancaster, Pennsylvania*

*Printed and bound by Berryville Graphics, Berryville, Virginia*

*Designed by Iris Weinstein*